# The Natural Religion

# The Natural Religion

Brendan Connolly

**EMMER PUBLICATIONS**

First published in Ireland in 2008 by Emmer Publications, Clonfane, Trim, County Meath, Ireland.

ISBN 978-0-9558313-0-0 (hardback)
ISBN 978-0-9558313-1-7 (paperback)

Copy-edited by Graham Russel
Typeset in 9½ on 13 point Plantin by Garfield Morgan
Printed in Great Britain by Lightning Source
Cover designer James Watson

# Acknowledgements

This book was written with the help of many people. First and foremost, I wish to thank those who gathered and communicated the knowledge that I used to form the ideas in this book. I feel fortunate and privileged to stand on the shoulders of these giants, excited at the heady altitude to which they have raised our knowledge.

I also wish to thank friends and colleagues, too many to mention, who over the years generously shared their ideas with me about the topics discussed in this book. In particular, the readers who so graciously gave of their time to read this book, gave me invaluable perspective on its content, and helped me immensely with making editorial decisions. Perhaps most important of all, their impressions supported my own conviction that this book had been worth writing. The readers were Harry Lloyd, Mike Murphy, Maeve Maclean, Edel Morrow, Gery Flynn, Linda Ward and Grainne Campbell. Any shortcomings of this book are entirely my responsibility and not that of the readers, and they do not necessarily agree with all the ideas in this book.

Finally, I wish to thank my mother who in raising me gave me the option to think beyond traditional boundaries, and also continuously provided sustenance during the research and writing of this book. I would like to thank Liz for her unfailing support, substantial editorial assistance, and endless patience.

**Brendan Connolly, B.A. mod., Ph.D.**

Brendan Connolly was born and attended primary school in The Netherlands. His family then moved to Ireland where he attended secondary school. He then completed a primary degree in Zoology, with Psychology as secondary subject, in Trinity College, Dublin; followed by a Ph.D. in Zoology on freshwater ecology in the National University of Ireland Galway. Subsequent work included Fishery Biologist with the Salmon Research Agency of Ireland and the Irish Central Fisheries Board. After this he specialised in Human Ecology and carried out research on fishing communities in Ireland and The Netherlands based in the Anthropology Department of Leiden University in The Netherlands. This book is the result of many years of preparation, followed by 3 years of full time research and one year of writing.

# Contents

# Why I Wrote this Book

Religions have a special role in the lives of people the world over, they address our personal and innermost feelings, and we have a unique sensitivity towards them. In recent times, millions of us have changed our attitudes towards religion. This has given rise to a variety of reactions, ranging from fear of a loss of morals and ethics, to a welcoming of the freedom to ask questions and form reasoned conclusions.

Living in our current knowledge revolution, and as a zoologist specializing in Human Ecology, I realized that I had information about which traditional religions only conjectured. As part of my work I focus on the ecological crunch-points of human life, such as the doubling not once but twice, of our world population during the 20th century. Since then, approximately 100 million people are being added to our world population each year, putting ever increasing pressure on food, clean water, and fossil fuel supplies. I also focus on the biological imperatives of our lives, like our personal relationships and the influence our emotional state of mind has on important decision making. In addition, my zoological training gave me information on what life is, how it works, how we humans originated, and why we die. This type of information may not normally be categorized as religious, but it is about explaining our human existence and our relationship with the universe around us; which is essentially also what the world's religions try to do. This is how the idea grew, that our current knowledge about ourselves could form the basis of a new, knowledge-based, religion.

Supernatural beliefs are part of most traditional religions. But beliefs differ and contradict each other and, being super-natural, cannot, by definition, be checked. Since I could not check which, if any, are correct, I could not use any supernatural beliefs as part of a new religion based on reasoning. For the same reason that supernatural ideas cannot be proven, they cannot be disproven either, because they don't apply to the natural world. Therefore, this new religion takes an agnostic standpoint, and is not atheistic because disproving something supernatural is as impossible as proving it. It is named The Natural

Religion, because it is not a supernatural religion. It is a fact-religion, not a faith-religion.

One advantage of using knowledge as a sounding board for a religion, is that knowledge is trans-cultural. Unlike supernatural beliefs, knowledge does not depend on where you are born. The Natural Religion can be used as a religion on its own, but because it is agnostic, it can also be used in conjunction with other religions. The knowledge of The Natural Religion can be used in everyday life, in combination with beliefs from faith-religions for those who also wish to have supernatural beliefs.

To put together this set of ideas as a religion was very compelling for me as researcher and author, because of the potential it has to contribute towards improving people's quality of life. Understanding human ecology improves our chances of success in managing human affairs. For example, poverty is a chronic shortage of resources and is, both nationally and internationally, a blight on human existence. Ecology, being the study of how species manage to make a living, is first and foremost concerned with availability of resources. By looking at the many ways in which other living species avoid running out of resources, we can learn how best to combat poverty amongst humans. Using knowledge as the basis for a religion reveals the practical reasons for our moral and ethical standards. This book explains how supernatural beliefs are not needed for morals and ethics, and that the value of idealism is the beneficial effect it can have on our everyday lives.

Religion makes idealism individually relevant, it is more deeply embedded in our psyche than philosophical and scientific knowledge. Religion affects us on a deeper, more personal, level. In this book I have tried to combine the impact that religion has, with the reliability of our new knowledge. This is a new direction in the transformations that religions have gone through over the course of time. The possibility that I, as one person and as a human ecologist, could help to improve the quality of people's lives the world over is truly exciting.

Brendan Connolly

# Chapter 1

# A New View: A New Religion

A new set of ideas

## 1.1: Questioning our existence

Understanding the meaning of life has been the quest of our human brain since its development. We can feel and think, we have emotions and we are self-aware, but what is the purpose of this? Why are we able to do this? A tree, for example, is also a living thing, but does not have emotions and is not self-aware like us. Yet, as far back in history as is known, people have wondered what it's all about, have tried to understand our own human existence, and tried explain the purpose of our own lives and searched for ways to help us make decisions. This urge to understand, this need for knowledge is at the centre of human nature.

We gather knowledge from many sources: popular wisdom, philosophies and the sciences have all been used to explore life's meaning. Religions, in particular, have traditionally played an important role in trying to explain our existence. Worldwide, cultures have formed religions, showing people's strong urge and need for a deeper understanding of ourselves and the world around us. While religions differ and have changed and come and gone, they are, in one form or another, a universal feature of human history.

In recent years, there has been a trend for people not to take traditional religions as seriously as we used to. Increasingly, people suspect dogmatic beliefs, preferring to reason things out and to understand, rather than to accept without question on blind faith. Also in recent years, we are experiencing a knowledge revolution; and as our knowledge continues to grow at an unprecedented rate, because more and more people have easy access to this information boom, we are now in a better position to question and reason than previous generations.

We now have the knowledge to answer such questions as: What is life? Where do we come from? Why are we here? As we will see, we don't need beliefs to address these anymore. Most religions were formed centuries ago, when our understanding and attitudes were very different. Because our knowledge has expanded so fast in the very recent past, it has in many respects moved far ahead of an outlook on life based on traditional religious views.

Changes in attitudes towards religion have led some to assume that we do not need religions anymore. However, as our human brain and consciousness have not changed appreciably during the last few millennia, we still have a need for the functions that religions try to fulfil. We still need knowledge and we still look for answers. The basic idea of a religion is to give us a set of answers to deep and personal questions and we continue to need these answers now just as generations of our forebears did in the past. However, answers based on outdated beliefs, myths and archaic thinking from many centuries ago are losing their credibility.

Religions are found the world over. Human cultures and societies – from the poles to the equator, and from the smallest island to the largest continent – have religions. So religions aim to fulfil a certain need in us all, but if we compare the actual beliefs of different religions, we see that they vary widely and even contradict each other. Some don't emphasize the rule of a supreme god (Confucianism); others have one god (Zoroastrianism, Judaism, Christianity, Islam); and still others have more than one god (Hinduism, Shinto). Some claim that their god or gods are the only true one(s) and that all other gods are mere superstitions (Christianity). Some religions believe that people have a supernatural soul (Christianity); others believe that we have several souls (Jewish Kabbalists and Hidatsa North American Indians). Others believe that people don't have souls at all (Buddhism), while still others believe that women don't have souls but men do (the Alawis of Syria). Belief in an afterlife can also vary from having one afterlife (Christianity and Islam) to having several by means of reincarnation (Hindu), or from having a final day of resurrection (Judaism) to the existence of a rebirth without transmigration (Buddhism).

In addition to all of this, there are myriads of other beliefs that differ from religion to religion. In fact, beliefs even differ within religions, resulting in different schisms, factions, sects and churches. Examples include Roman

Catholicism, the Eastern Orthodox Church and the many Protestant churches of the Christian religion; the Sunnis and Shi'as of Islam; and Buddhism's Theravada, Mayahana and offshoots such as Ch'an and Zen. While humans all over the world have religions, beliefs differ both within and between the world's religions.

## 1.2: What do religions have in common?

Religions differ, but they also have certain things in common. They address questions about our existence, our life and death and the world around us. They advise and have rules about making ethical decisions and how we should balance our own interests with those of others. Most religions tend to focus on us personally and how we deal with the world around us. Religions tell us about what we are, what we should do, and where we are going. The answers to these questions are of deep importance to us. Religions are a psychological and emotional base for billions of people.

Why is there now a decline in interest in religions when people from all over the world have always had them? What has changed? Are people not interested anymore in the age-old questions about the meaning of life? Or are we losing faith in supernatural beliefs?

A change that has taken place in recent times has been our vast increase in knowledge. Scientists, philosophers and other academics dedicate entire lifetimes of work to gathering knowledge. The effort the human species is putting into understanding life has surged ahead. This shows that people are as interested as ever in the meaning of our lives, but are looking less and less to supernatural beliefs for answers. Our new knowledge also gives us a broader perspective on specific supernatural beliefs from around the world and it tells us how different and contradictory these beliefs are. Knowledge has put religions in a whole new light and has changed people's attitudes towards them, particularly in the richer regions of the world where people have more education and more knowledge at their disposal.

The information boom has not taken away people's need for a solid base for our emotional and psychological state of mind. The worldwide existence of religions is proof enough of that. We still look for the answers that many religions have traditionally tried to give, even if we have lost faith in supernatural beliefs. People need to make practical decisions on a daily basis. For example, we need to help others, yet we also need to care for ourselves and at times these may conflict. So, how do we reconcile the two? In a wider context, we need codes of ethics for reasons that have nothing to do with anything supernatural, but are needed for practical reasons such as organizing our societies. A sense of meaning and usefulness in our lives is important for our own emotional well-being while we are alive, not for after we stop living.

Looking for answers to important questions is something that religions have in common. Many religions use the promise of supernatural rewards and the threat of supernatural punishments as reasons for making decisions in our lives. But our decisions in dealing with practical problems should be influenced by the real and actual consequences, not hypothetical supernatural consequences.

## 1.3: Religions in the past

The first priorities for most of us are to stay alive and to breathe, drink, eat, have shelter, be safe and enjoy companionship; but we also have an inborn need to understand ourselves and the world around us. In the past, people guessed the answers to these fundamental questions because we did not have the knowledge to answer them in any other way. Since these answers were only guesses and since there was no information to back them up, these answers could only be accepted on faith. Thus whole systems of beliefs emerged based on blind faith. To have faith in a belief means that you accept that belief as fact without having any evidence to support this. Many questions were answered by religions with beliefs held purely on faith.

As most people didn't have the time or the resources to fully research and investigate the meaning of life, religions presented a collection of explanations and ideas in one integrated package. This package was offered to us as one unit, a religion, and gave people a set of answers that attempted to satisfy our human need to know.

## 1.4: Religion of the future

This book is about our search for answers to the most important personal questions anyone can ever ask. This need to understand everything is an essential part of who we are; as a species we have evolved this incessant curiosity. It is our knowledge that most distinguishes human ecology from the ecology of other species and it has given us extraordinary advantages and power compared to any other animal. This thirst for knowledge also drives our need to understand the reason for and the meaning of our own existence.

**These ten chapters offer an integrated set of ideas about our lives, just like religions have traditionally done, but with the important difference that no supernatural beliefs of any sort are used. This set of ideas**

**is exclusively based on up-to-date knowledge; it is a discussion about the meaning of our lives, and the implications of our new knowledge for our lives.**

The reason we can now take this new approach is that, at our present stage in human development, we have discovered many of the answers that we could only guess at in the past. We don't necessarily need beliefs as our psychological base and we don't need to guess at the meaning of our lives. We can now find answers to those important personal questions in our knowledge.

It is argued by some that religions are not needed anymore. However, judging from the role that religions have played all over the world this seems unlikely. People's *psychological* needs cannot have changed that much during the last two or three millennia. This book hopes to close the gap between a world view based on traditional religions, and our knowledge boom which has changed so much and moved so far ahead in comparison. It hopes to show the implications of our new knowledge for our view of ourselves and the world around us.

Religion in the future needs to develop and improve on religions of the past because in the past many problems remained unsolved. In addition, times have changed since traditional religions developed and so new solutions are needed. Problems that humankind continues to face range from intimate personal dilemmas to difficulties on an international scale. For example, taking a broad overview, there are now more than 6.6 billion people on Earth, while two millennia ago the human population is estimated to have been between 130 and 250 millions worldwide – just 2% to 4% of what it is now. This huge increase in population has created very different and greater practical problems than those facing us 2,000 years ago. Such complex problems need to be addressed using the most powerful tools at our disposal. To continue to address such intractable problems using unverifiable beliefs, while verifiable knowledge is available, is nothing short of irresponsible.

Religions have great influence on people's lives, so for the future we need to improve religions as much as we can. Knowledge can be used directly by those who manage our societies such as economic planners, politicians and leaders of our communities. Knowledge also helps each one of us as individuals to understand ourselves and the world around us. The function of a religion is to explain our lives, our psychological state of mind, and to put ourselves in context with the rest of the world.

The way religion in the future can improve on religions in the past, is to make the connection between our new knowledge and our innate mental needs.

The next chapters offer a new alternative, combining our modern knowledge with our primeval need for answers, without using out-of-date and unsatisfying beliefs and dogmas. We can use our knowledge as the basis for a new religion and address the important questions and give the personal support that we need. Belief in supernatural ideas doesn't need to be part of that support. Because the answers and ideas of this new type of religion are based on knowledge, they are more likely to be correct and decisions based on them are likely to be more successful. One reason for this is that ideas and suggestions based on knowledge can be verified, unlike ideas based on supernatural beliefs. The emotional and psychological base of this new religion is therefore more solid, more realistic, and more reliable than that of a religion based on supernatural beliefs.

The hope is to guide the progression in the development of religion towards knowledge and reasoning, and away from dogmatic and supernatural beliefs. It is the ideas of this new view of religion that are most important. It has been given a name, but this is solely to help describe it more easily and is not to be used for any type of veneration as with other religions. The new approach suggested here is named "The Natural Religion".

## 1.5: The Natural Religion – a personal choice

In recent times, the freedom to choose religious affiliation has been increasing in many societies worldwide. For more and more people, it is now possible to examine a range of religions and to choose one, or several, or different aspects from different religions, or none at all. Choice of religion is a very personal matter. Which religion we choose depends on where we were born, the culture and time in which we live, and our individual personality.

Knowledge applies to everyone and is therefore not like religion. Humankind strives for facts that are the same for every person, no matter where we are born or in which culture or community we live, but – unlike knowledge – selecting a religion should always be a personal choice. **One of the aims of The Natural Religion is to contribute towards giving people the option of an informed choice of religion.**

The following chapters explain the alternative and new approach of The Natural Religion.

(As the terms **'knowledge information'** and **'genetic information'**, as well as **'fact'** and **'faith'** have special significance in this book, they are fully discussed and compared in the Appendix. For a full appreciation of the ideas in the following chapters, it is advisable to read the Appendix now.)

# Chapter 2

# Life

Planet Earth has spawned life, we are part of that

## 2.1: Life thinking about life

What is life? What is the difference between the rock, clay and water that make up our planet, and the living beings that inhabit it? And is our own human life the same as that of other living things? These are some of the questions which religions have tried to answer since time immemorial.

We humans live and think about the fact that we live. We associate our life with feeling and thinking and being aware of our own existence. Our brain is part of life and has thoughts in the form of bio-electrochemical models (see Appendix) about the fact that it has thoughts – bio-electrochemical models modelling themselves.

In the 19th and 20th centuries we have become aware of how our life processes work. Right now, in the absence of evidence to the contrary, it would seem that we humans are the only life form on Earth that has acquired an understanding of our own life.

## 2.2: **The origin and essence of life**

Human beings are part of life, and everything we create is a product of life, but what makes life different from non-living material? What defines life?

One of the characteristics of life is that most living things grow. However, crystals, such as ice, can also grow, but this is generally not seen as life. Another characteristic of life is that it reproduces. However, certain types of chemical molecules, such as clay crystals, form images and mirror images of themselves, which essentially is also reproduction. Thinking and feeling, which we humans do, is part of life, but it is not part of all life. Bacteria and microscopic plants, for example, do not have nerve cells or brains to carry feelings or thoughts, yet we still see them as being alive.

Up to the 20th century, the essence and origin of life was commonly explained in supernatural terms. The theory of spontaneous generation of life – that life can arise from nothing – was suggested by ancient Greek philosophers. Up to the latter half of the 1800s some thought that life was somehow eternal and had no beginning. However, others during this time began to wonder whether life had begun from a 'pool' of chemicals in certain appropriate conditions of temperature, light and electricity. It wasn't until the 1920s that specific theories were formulated stating that life could have started from simple non-living constituents.

A number of experiments were carried out in the 1950s which reported that if you put methane, hydrogen, ammonia and steam in a container in which there is also an electrical discharge, then amino acids (basic sub-units of a protein) are produced. Amino acids and proteins are very important chemicals because they are constituents of living cells and are classed as biochemicals. Compared to the simpler chemicals from which they are made, they are very large and complex and are sometimes called biomacromolecules. To give an idea of the difference in size between a simple molecule of salt or water and biochemical molecules, if a water molecule were the size of a football then a biochemical molecule could be as big as a house. Proteins are crucial in the process of life and are made from non-living chemicals. These proteins, on their own, are not full living cells, but they are an essential part of cells. Since then, more experiments have produced other basic building blocks of living tissue. For instance, complex carbohydrates have also been made from simple chemicals.

Currently on Earth, large and complex biochemical molecules generally do not exist for long outside living beings, they need special conditions such as a laboratory to exist on their own. If they do occur, as in the case of a dead and decomposing plant or animal, they are rapidly taken up by other living organisms such as bacteria and larger scavengers. Most living beings shield their tissues and their large complex biochemical molecules from the outside with a protective skin.

Life appears to have developed from a simple beginning and has evolved and changed all the time, and continues to do so. Before life first appeared on Earth, the Earth's atmosphere was not the same as it is now. Amongst other differences, it is generally accepted that at that time there was little or no oxygen in the atmosphere. One theory proposes that this primordial atmosphere allowed chemicals to combine in what is described as chemical evolution, which then formed the biochemical building blocks of living cells and tissues. With no living bacteria or other primitive life forms around to mop them up as would happen today, these large biochemical molecules could have existed in the open, drifting around and combining in various ways. Subsequently, according to this theory, a second type of biochemical evolution could have taken place, in which these biochemicals combined to form larger globules or types of bubbles. These bubbles could then have taken other bubbles inside them to result in what is essentially the basic structure of a living cell. This is one theory about the origin of life.

Other theories exist. One suggests that meteorites crashing into Earth possibly carried some sort of primitive life with them. This theory presupposes that living cells survived the journey and its extreme temperatures – perhaps in a resistant state similar to bacterial spores or seeds – thus introducing life to planet Earth. If life did not develop here, but 'seeded' Earth from somewhere else, it would still have had to develop from nonliving materials elsewhere. Amino acids and sugars have been found on a type of meteorite called carbonaceous chondrites that have crashed into Earth, so we now know that these building blocks of proteins have also formed elsewhere in the Universe.

Life, as we know it, can only exist in a certain narrow temperature range, from about zero degrees Celsius (32 degrees Fahrenheit) to a maximum of between 40 and 50 degrees Celsius (104–122 degrees Fahrenheit) for most of life. We assume that if life developed or exists on another planet it would also have to be within a certain temperature range. Some commentators have dubbed this the 'Goldilocks range': not too hot, not too cold, but 'just right'.

From what we currently know about life we can draw two conclusions. First, life is a highly complex combination of chemical compounds. These compounds are made up of ordinary chemicals which also exist outside of life. No other agent, spark, spirit or life-force is necessary for life. The minute complexity of both the biochemistry and the processes of life are staggering. **The difference between life and non-living things lies in the vastly more**

**complex way in which chemicals and biochemicals are combined, and react with each other, within living organisms than they do outside of life.** Essentially life is due to the complexity with which chemicals combine and react with each other. Life's complexity and the fact it came about is truly awe-inspiring. **Nevertheless the indications are that the processes which developed into life are chemical and not supernatural, and therefore this idea has also been included in The Natural Religion.**

The second conclusion is that all life that we know about, including human life, is fundamentally made up in the same way. **This same basic plan of construction and manner of operation suggests that all life that has survived on Earth probably originated from a common beginning.** This means that if we go back far enough in time, all presently living species – including plants, animals, microbes and ourselves – have a common ancestry and are therefore related.

It is relatively easy to visualize that we humans have a common ancestor with the great apes because we look like them. Estimates of the genetic similarity between ourselves and common chimpanzees are reported to show over 98% similarity, and with pigmy chimpanzees or bonobos there is over 99% genetic similarity. To put this in perspective, the genetic similarity between humans and chickens is estimated to be over 70%. After all we do have four limbs, a head and a body with various organs in it and so do chickens, so the basic plan is the same. On the other hand, it is not as easy to visualize our common ancestor with, say, grass, but we do have gene sequences in common indicating that we do indeed have a common ancestor with grass. However, our common ancestor with grass lived a much longer time ago than our common ancestor with the great apes. Therefore part of The Natural Religion is that all life on Earth is connected, and most likely descended from a common ancestry.

## 2.3: Creating life

Living organisms, including humans, reproduce. Life recreates new life which is similar to it by a process of reproduction. As life is composed of a complex of chemicals, the question arises whether people could make a living organism from non-living materials and chemicals. If our knowledge about the bio-chemical processes involved in life continues to grow the way it did during the last 100 years or so, then yes, it does seem likely that at some stage in the future we will be able to create a living being directly from non-living materials.

Fears have been expressed that man's imminent knowledge and capacity to create life from non-living material could have negative consequences. These fears centre on the possibility of making life that could be used in a destructive way.

We have already changed disease organisms from other living animals so that they will attack people. Humans do have the inclination to produce lethal biological organisms, is it therefore necessary to forbid people to work towards creating life from non-living materials?

If we forbid research in the general area of the operation of life, we would have to forbid a huge area of work. Much of this type of research is quite crucial as its results are of immediate practical benefit to areas such as agriculture, food storage and preservation, and medicine. Biochemical and biomedical research also gives us important information about ourselves and the way life functions, such as which diet is best for us, how to strengthen our immune systems and how to use antibiotics. Different areas of research are often very closely linked and advances in one area of biochemistry often help progress in other areas. For example, research into amino acids gives us information about a series of other larger biochemicals because they are the sub-units of proteins. Research into the large molecules that make up proteins and fats gives us information of how the membrane that encloses a living cell functions. Understanding the properties of a cell membrane not only tells us how the organs of our bodies operate, but is also used in biotechnical industries. In addition to this, new and improved research techniques discovered in one area can often be of great benefit in other areas of research. DNA typing is of great benefit to medicine but is now also used to make crime investigations more reliable.

Our human ecology is very much dependent on our accumulation and use of knowledge information. To retard or stop the gaining of knowledge would be a retrogressive step. This has happened in the past. One example of this is the rise and decline of China. China in the medieval period was technologically and culturally far advanced compared to other countries in the world. However, at a certain stage in its history, China tried to close itself off from outside influences and stopped progressing with the rest of the world. This resulted in other regions in the world overtaking China, from whose culture they had learned so much in the first place. Learning and advances in knowledge have always been the hallmark of a strong and successful culture or society. As we saw before, losing knowledge or retarding its growth can be detrimental and cause stagnation and decline in a society. In human ecology, advancing knowledge has always been central to our success and has given us greater power than any other species.

A ban on research into creating a living cell could have unintended repercussions, such as clandestine operations working outside the law. Unmonitored and uncontrolled research increases the potential for mistakes in, and misuse of, that research. Therefore, included in The Natural Religion is that forbidding biochemical and biomedical research into the possible creation of life would be retrogressive and could be counterproductive. In order to control the possible misuse of this type of research, open and transparent monitoring is as necessary as it is in many other areas of human activity.

## 2.4: The function of life

Why did this highly complicated chemical activity called life develop? The answer is energy.

The Sun's rays travel through space and a small percentage of them hit Earth. These rays consist of energy. The living system on Earth traps some of this energy and uses it to breathe, eat, grow, reproduce, move and do all the other things that living things do – including reading! The way in which the living system traps, stores, and uses this energy is with and in large bio-chemical molecules and compounds.

As we saw, at the time that life began, certain chemical molecules had formed and were freely floating around. Some of these molecules contained chemical bonds that trapped and stored energy. From then until the present day, circumstances on Earth have allowed these biochemical molecules to grow larger and larger and to grab and hold on to more and more energy. Life uses energy to grow so it can then gather and store more energy. This process happened during our human evolution, but also takes place during our individual growth from conception, through infancy, and then on to childhood and adulthood.

All animals depend either directly or indirectly on plants to trap the Sun's energy for them. Without plants trapping this energy and storing it in their tissues, the plant-eating animals could not live, and neither could their predators. For example: when we eat tuna, that tuna could have eaten a mackerel, that ate a sprat, that ate zoo plankton, that ate algal plankton. At some stage in this food chain one animal has to have eaten a plant, which in this case is algal plankton, because animals cannot trap energy directly from the Sun – only plants can do this. Animals depend on plants for their everyday food. If plants stopped growing, animals, including ourselves, would starve to death. We humans eat plants as well as animals, so we get the Sun's energy both directly and indirectly from plants.

The reason plants can trap the Sun's energy is because of the process of photosynthesis. This process which occurs in the green parts of a plant is a complicated series of biochemical reactions in which sunlight is taken in and locked into chemical bonds. The vast majority of energy that circulates through, and powers, all life on Earth, is locked in from the Sun's rays by photosynthesis.

There are forms of life that live in the depths of the oceans where there is no sunlight. However, bacteria can grow down there which allow other species to also live there which feed on them. It is via these bacteria that energy gets into the food chain, but rather than getting energy from the Sun these bacteria get energy from certain chemicals which come out of volcanic vents in the ocean floor. They break these chemicals apart, taking the energy that is stored in their bonds and incorporate it into biochemical bonds within their own cell bodies.

So, here also the process of life consists of gathering and storing energy from one set of chemical bonds to another all the way up the food chain. However, the vast majority of life on Earth gets its energy from the Sun.

The Sun is one of over 100 billion stars in our galaxy, the Milky Way. Our galaxy is only one of many galaxies in the vast and still expanding universe. The energy for most of life on Earth comes from our Sun. Not all stars are the same temperature as the Sun, nor are they the same size. Conditions on Earth have obviously allowed life to develop, but if our Sun had been hotter or colder or had burned out before now, then life might not have had the conditions to develop. Life needed time to develop; evolution takes place over time, so without time nothing can happen irrespective of other conditions. If the Sun had burned faster, life would either not have developed to the stage at which it is now or would not have developed at all. So not only did the particular circumstances on Earth allow life to develop, the particular type of star which the Sun is has also been crucial.

The Sun is approximately 4,500 million (4.5 billion) years old, and it has enough fuel to burn for approximately another 5,000 million years. When the Sun runs out of fuel, our whole solar system will change irrevocably and the Earth will be destroyed. Planet Earth at the moment is neither too hot like Venus (which is the planet closest to us, between us and the Sun), nor too cold like Mars (on the other side of us). Our Moon has a steadying effect on the way the Earth moves in its orbit, but the Moon is imperceptibly moving away from Earth by 4 centimetres per year, so in approximately 2,000 million years it will have moved away so far that the Earth's motion will change considerably. These periods of time are so long that we as humans find them impossible to comprehend. 2,000 million years theoretically represents about 66 million human generations and in 5,000 million years we could have 166 million human generations, but in the future these events will happen in our solar system.

Other catastrophic events can also happen in the short term. Such as a large meteor hitting Earth or a supervolcano eruption such as is theoretically expected for Yellowstone Park in the USA's Rocky Mountains. The gigantic dust clouds sent into the atmosphere as a result of either such an event could block out sunlight to an extent that conditions on Earth could be beyond what we and many other animals and plants need to survive. These are all natural cataclysms which we currently cannot do anything about, in contrast to other catastrophes such as wars fought using nuclear weapons that we can avoid.

Our own existence as a species so far has been estimated to be approximately 100,000 years old. Will we still exist in 5,000 million years when the Sun burns out? No, we will not exist in the form we have now. We will have become extinct or we will have changed genetically into a series of other species long before then. Earth has changed during the last 4,500 million years and we have to presume it will continue to change for the next 5,000 million years. Any one of a number of changes, such as meteor impact, a supervolcano eruption, or major climate change due to a build-up of hothouse gases in the atmosphere, could cause a rise or fall in temperature that could bring about our extinction long before then.

One way we may be able to prolong our own existence as a species, is by means of our knowledge information. A combination of technology and genetic know-how may be able to either shelter us from, or adapt us to, Earth's changing environment. We could leave it up to the genetic information in our genes to change in the traditional way, as it has done in our ancestors and those of all other species of living organisms on Earth. However, we then have to assume that we will either become extinct or change out of all recognition, as this is what has happened with all species that ever existed on Earth. On the other hand, as far as we know, no other species has ever developed the capacity to use knowledge information to the extent that we have. This is why the ideas which form The Natural Religion include that using our knowledge information we could potentially prolong our existence as a species. Whether this is the case and for how long we can prolong our species' existence, depends on whether we continue to gather and develop our knowledge and how we then use it.

There are, of course, many other sources of energy in the universe such as other suns and other energy rays. We may in the future be able to access and use these either from the Earth or by leaving Earth and escaping our solar system, when either our Sun or Earth's orbit around it becomes unstable. All this will depend on whether we can gather the necessary knowledge to enable us to do so. After all, there is time for 60 million human generations before Earth's orbit will change and over 100 million human generations before the Sun becomes unstable. In addition to this, we will also need more knowledge if we are to have even a small chance of surviving the other changes that will occur on Earth in the meantime.

Species that we know from fossil records that lasted over millions of years, went through a period of emergence, followed by a blossoming, after which the species died away or changed to another species. We have to expect that we, as a species, will follow the same path of development first then widespread growth followed by decline or change. Using our knowledge could improve our situation, but it could also make it worse. Atom bomb explosions killing people and causing nuclear winters, overpopulation hand in hand with food shortages, climate change due to our pollution of the atmosphere, or diseases engineered by us for biological warfare, are all ways in which our knowledge could seriously threaten our existence. Knowledge information is very powerful either for good or for bad.

**Our position as human beings, living in our solar system on planet Earth, is therefore that we get our energy for breathing, growing, feeling, thinking – for everything we do – from the Sun. We cannot get energy from the Sun directly, we need plants to trap the Sun's energy for us. Without plants we cannot live, therefore there are very practical and compelling reasons for us to respect other living species that share the planet with us.**

The fundamental purpose of human life is the same as that of all other life. We develop and grow in order to gather energy, so that we can reproduce and enable our children to gather energy, so they too can reproduce in their turn. Our genes have survived via generations of our forebears, and will continue to attempt to survive and reproduce via our current generation and the next generations. For these genes to successfully survive and continue their journey through the generations, we as individuals need to survive and prosper and nurture our children. Life's strategy is to maximize its quality of life and chances for survival.

Religions have traditionally aimed to contribute towards improving our psychological welfare and quality of life, and in doing so try to help us have the energy to carry out the main function of our lives. Improving the quality of our life and psychological welfare is also the main reason why The Natural Religion is introduced as an alternative concept, but it does this by using up-to-date knowledge as its base and thereby gaining more reliability than offered by the older perspectives of traditional religions.

## 2.5: Everything is energy

The process of living consists of highly complex reactions of very large chemical compounds. Any chemical compound is made of atoms which can belong to different elements. Which element an atom belongs to is determined by the size of the atom. For example hydrogen, carbon, oxygen, or iron are all elements, and their atoms are different in size. If a chemical has atoms belonging to different elements in it, it is a compound. Table salt is a chemical compound because it is made up of one atom each of the elements of sodium and chlorine bound together to form a molecule of salt. Each of the elements have different chemical characteristics which cause their atoms to combine in different combinations. These combinations make up the compounds that are the chemicals of life. As much as 96% of the human body is made up of the atoms of just four elements combined in many different and complex ways. These elements are carbon, oxygen, hydrogen and nitrogen.

Atoms are all formed according to the same basic plan. They have a small area at the centre called the nucleus which can contain two types of sub-atomic particles, protons and neutrons. Protons have a positive electrical charge and neutrons have no electrical charge. The protons and neutrons both have weight. The combined number of them in the nucleus determines the weight of the atom, and therefore also which element that atom belongs to. For example, the atomic weight of carbon is 12, for oxygen it is 16, hydrogen is 1 and nitrogen is 14.

Electrons spin around outside the nucleus. These electrons have negligible weight, but they have a negative electrical charge. In the larger atoms, there may be many electrons. These electrons tend to be distributed in layers or orbits around the nucleus, some close in and some further away. They form a sort of electron-cloud around the nucleus. Normally the negative charge of the electrons balance the positive charge of the protons. Depending on how many electrons and protons an atom has, the outside electrons can spin off away from the nucleus. If this happens, the whole atom has a positive charge because it lost a negatively charged electron. The alternative also happens. An extra electron can join the electron-cloud and this then gives the whole atom a negative charge.

The number of protons and neutrons and the number of electrons which are likely to either join or leave the atom gives each element its particular chemical characteristics. If one atom has a positive charge and another has a negative charge, they are likely to join up to form a bigger molecule of a new compound such as a molecule of table salt. In this way all chemical compounds, including the ones involved in life, are built up.

In the early and middle 1800s, atoms were thought to be the smallest particles that existed. Electrons were discovered by the late 1800s, and protons and neutrons during the early half of the 20th century. Then for a number of years protons, neutrons and electrons were thought to be the smallest particles that existed. However, physics experiments in the 1960s indicated that the protons and neutrons in the nucleus consist of even smaller parts.

The sub-parts of protons and neutrons are called quarks which appear to be held together by other particles called gluons. It appears there are as many as six different types of quarks. The different types are sometimes called different 'flavours'. Each proton and each neutron contains three quarks. A proton appears to have two 'up' quarks and one 'down' quark, while a neutron has two 'down' quarks and one 'up' quark.

If we then ask what quarks are made of, the current answer is that no internal structure is known for quarks. Quarks are sometimes described as packages of energy. At the level of quarks, we are getting close to the energy—matter interface. All matter is made up of energy and when we examine quarks we are either at, or very close to, energy.

Research in physics is continuing to, both theoretically and actually, identify further particles and packages of energy. Concepts which are likened to vibrating strings and planes, and numbers of dimensions running into the 20s, are used to described how energy forms matter, how gravity works and how time relates to all of this especially before the 'Big Bang' occurred. While all these new ideas and concepts are extremely difficult to understand, and while some of these concepts as far as we currently know are not at all certain to actually exist, **it does seem that energy is the most fundamental entity of which everything else including ourselves is made.**

Based on the knowledge which is currently available to us, the following overview is included in The Natural Religion. Starting from energy there is a continuum all the way to living organisms such as us. Part of this continuum are all the characteristics of human life, including our mental capabilities such as our feelings and emotions, our self-awareness and our knowledge and all our psychological states and abilities. If we were to go back along the line from our bodies to our organs, bones, muscles, to the tissues and the cells of which they are made, we would eventually end up with energy. The following list starts at energy and indicates the stages between energy and us and our human characteristics:

- **Energy** is what makes up **quarks**;
- **quarks** are in **sub-atomic particles**;
- **sub-atomic particles** make up **atoms**;
- **atoms** make up **chemical elements**;
- **chemical elements** make up **chemical compounds**;
- **chemical compounds** make up **biochemical compounds**;
- **biochemical compounds** make up **sub-cellular organelles and structures** (e.g.: nucleus, mitochondria, cell membrane);

- **sub-cellular organelles and structures** make up **living cells;**
- **living cells** make up **living tissue and organs;**
- **living tissue and organs** make up **plants and animals;**
- **animal life including human life** produces **feelings, emotions, knowledge, self-awareness.**

This continuum – from energy all the way to us humans – is what we currently consider to be correct. The reason why we have such confidence in this interpretation is that it agrees with observations ranging from mathematical calculations about energy, experiments in physics on quarks and sub-atomic particles, experiments in chemistry on the biochemistry of cells, to phenomena in the biological world of which we ourselves are a part. The continuum from energy to ourselves shows where we fit into the world around us, it puts us in context in the universe and summarizes what we are made of.

In many cultures past and present, life is seen as separate and apart from the non-living world, and people are seen as separate and apart from all other life. This view is not supported by our current knowledge of how the universe, our solar system, and life on Earth – including our own – functions. In fact there is a connection through energy between everything in the universe. **We are part of the universe and are made of the same energy as the rest of the universe. Energy's particular properties give rise to the chemical structures that are the foundation of our shape and abilities. The conclusions included in The Natural Religion are therefore, that since energy can be observed and measured and has physical properties which can be examined and tested, it is not a spirit, an unexplained presence, or supernatural in any way.**

## 2.6: Energy – the limit of our knowledge

Everything in the world including ourselves is, at its most basic level, energy. We simply do not know what lies beyond energy. Our knowledge information has limits and this is one of them. We should attempt to push this limit further back, but it could be the case that there will be limits to our knowledge for a long time to come and maybe for always.

The idea that there are things in our universe which we don't understand or simply don't know about, can be unsettling to the human consciousness. Traditionally, gods, the mystical and the supernatural, occupied the realm beyond the limits of our knowledge – the unknown. Our inbuilt urge is to understand everything around us. Knowledge is power and gives us a feeling of safety, and enables us to control our lives. Having mystical and supernatural explanations for what we don't know, reduces our apprehension of the unknown and many people are satisfied by this. However, it is more realistic to recognize the limits of our knowledge information. Realism gives the knowledge we have more reliability, and therefore decisions based on realistic knowledge will be more successful. Moreover, accepting a lack of knowledge can be more satisfying than guessing at explanations, particularly when guessed explanations differ and contradict each other and cannot be verified in any way. Beliefs in supernatural concepts and ideas are actually guessed explanations.

There is a further beneficial consequence from accepting that we don't know certain things, which is that it focuses our attention on that gap in our knowledge, motivating us to fill this gap. The more energy we devote to examining and enquiring into things, the more likely it is that we will gain extra knowledge. If we have beliefs held on faith we tend not to question them because that would mean that we might not really believe in them. In fact, inquiry into beliefs used to be, and in some cases still is, actively discouraged by many traditional religions. Therefore, adopting beliefs about things that are beyond the limit of our knowledge, can have a retarding effect on expanding our knowledge. There are many examples of this throughout human history, one famous one in the early 1600s was Galileo Galilei's lifelong struggle against restrictions on free inquiry, culminating in his being brought before Christianity's Roman Catholic Church's Inquisition and his subsequent house arrest.

**We humans have to take decisions about our future, and these are more likely to be successful if they are based on realistic knowledge. This is why the position of The Natural Religion is that even though our knowledge has limits, our human knowledge should be taken more seriously than beliefs held on faith.**

## 2.7: The system of life – why does it exist?

To the question: 'which came first, the chicken or the egg?', the only correct answer is that the Earth came first. Planet Earth most likely originated approximately 4,500 million years ago as a result of the 'Big Bang' which occurred billions of years previously to that again. At first the Earth appears to have been extremely hot and molten, and also smaller than it is now. The Earth most likely grew by bumping into other objects in its orbit including comets made of ice that then became part of the Earth. The whole planet then cooled and the outside solidified. The ice comets melted and formed water on the surface and only then did conditions in the water and atmosphere and on land allow first life's biochemical precursors and then life itself to develop.

The Earth spawned life. It is estimated that life emerged on Earth between 2,800 and 3,500 million years ago. As we saw, the enormous biochemical process called life can only function because conditions on Earth allow it. The reason why life exists on Earth is because of the type of planet it is. **Our system of life is part of, and inextricably linked to, the Earth. It is one of the chemical processes which make up this planet. Life is as much an interactive product of Earth as leaves are of a tree. Life is a breathing, growing, reproducing, moving, feeling, thinking part of Earth's surface.**

Each individual living being is intertwined with its living and non-living environment. Some illustrate the closely interlinked nature of life and our planet by saying that the whole planet is alive, but it is perhaps more accurate to describe planet Earth as a huge chemical complex of which life is one part. Most of this planet is usually defined as non-living, but the fact that life is more complex and the non-living part of Earth is less complex does not mean that they are qualitatively different. Both the living and the non-living are chemical reactions.

As life is part of what planet Earth is, it partly produces the circumstances that exist on Earth and it has done so ever since life on Earth began. For example, in the beginning oxygen was either not in the Earth's atmosphere or was only present in very small quantities. Yet now it makes up 21% of our atmosphere. It is thought that most of the oxygen in the air came either directly or indirectly from photosynthesis in plants. Because oxygen is being maintained at 21% at the moment and given that the majority of living species use oxygen to live, it is safe to conclude that a proportion of life on Earth is returning oxygen to the atmosphere. The amount of oxygen used up and returned each year to our atmosphere is estimated at 100,000 megatons. Oxygen may be a non-biological part of our environment, but most of it is actually a product of life. Likewise many other gases in our atmosphere are kept in balance, most likely as the result of life in our oceans and on land.

There are of course many other factors that influence us in our environment that themselves are part of life. As animals, we would starve to death if plants stopped producing food for us. Diseases and parasites such as the black death, cholera, small pox, tuberculosis, and malaria are all part of life and have a big influence on us as they are historically the biggest killers of people.

We humans are part of life, and have also greatly influenced circumstances on Earth in our turn in many ways. We have affected the ecology of many plants and animals including of course our own ecology. We have had this effect particularly through using our knowledge, which is, as we already saw, itself a product of life. Our agriculture has transformed huge swathes of the Earth's surface. We have constructed a new type of human habitat by creating urban areas. Our air pollution affects many living organisms including ourselves by influencing our climate, and we have created arid land by removing vegetation. We have made land out of the sea, transformed some rivers into lakes, and sucked other rivers dry of all their water. We may yet be able to leave planet Earth and start life and new ecological systems on other planets.

Ecology is the study of the way Earth allows life to exist, it examines life's interactions with everything around it, including other life. If we look at life and the way it changes and survives, we first have to examine its ecology.

Humans have an ecology like all other living beings. We prey on other plants and animals for food, we need a place to live, we reproduce, we get diseases and parasites, and we compete with each other and other species. These are the basics of the ecology of any living species, whether you are an earthworm, grass, a bacterium, a giraffe, a beech tree or a human being. Ecology formed and drove evolution, not as it is sometimes portrayed, the other way around. Evolution is ecology's and life's way of getting around the problem of the earthly environment changing all the time.

'Evolution by means of Natural Selection' is the process by which life changes from one generation to the next. Sexual reproduction results in off-spring which are to a greater or lesser extent genetically different from their parents as well as from each other. Depending on circumstances, some of these may be more successful or better adapted than their siblings or their parents. So there will be a growing proportion of the more successful individuals in future generations. Why has this system of inbuilt change developed? Why go to all the trouble of having a variety of types of individuals in a population? The reason is that the earth's environment changes all the time. Initially the whole planet cooled down from a scorching molten mass and went through a whole process of formation of land and sea. The Earth changed from one ice age to the next. It is still changing because the world's continents are slowly drifting towards and away from each other. Weather patterns vary continuously over time. Of course, as mentioned above, living species themselves also change the Earth's environment, and food supplies change as part of regular ecological cycles. It is estimated that during the course of evolution 99% of the species that ever existed on Earth are either extinct or have evolved into another species. This gives some indication of how much the Earth's environment has constantly changed. The environment on Earth has never stayed the same in the past, so we have to assume it will continue to change in the future. Evolution anticipates the changing circumstances on Earth by producing a variety of types of offspring. Depending on how much and how fast the environment changes, some find it easier to live and compete for resources than others. The changing environment is one very basic driving force in the ecology of all living organisms and is also the reason why individual living beings die. Our own human mortality and its implications are discussed in the next chapter.

We humans are the result of thousands of millions of years of life. Each generation was slightly different from the previous one, and then died after reproducing. The changes in hundreds of millions of generations accumulated, and resulted in a series of different species, one evolving from the next, of which we are the most recent. Through time, life has accumulated a vast and complex system and a huge store of genetic information (see Appendix). On average, the additional new amount of genetic information which we gain at each new generation is absolutely minute compared to the amount of accumulated genetic information which we inherit from our parents, but these minute changes have over millions of generations of our predecessors accumulated to form us.

Religions have always theorized about our origins, the meaning of our lives, and why we are what we are. From our new and better understanding of the way life works, we now have realistic answers to these questions. The reason we are now alive is because of our ancestors' efforts to survive and reproduce. These ancestors belonged to a long series of species that existed over several thousands of millions of years. **The meaning of our ancestors' lives is the fact that we exist, and likewise the meaning of our lives is what we contribute to the future of humankind and any new species into which we may yet develop.** These answers and interpretations of the basic meaning of our existence are based on our current knowledge and are part of the new development in religion offered here.

When we look at an ecological system such as a habitat with all its microbes, its animals, and its plants, we see a situation at a particular point in time. In many of our Earth's great ecological systems, of which the seas are the biggest, it seems as if everything was formed just to link in and interlock with the other living species of that ecological system. The life cycles and movements of the plankton in the seas seem to be perfectly coordinated with the needs of the animals that eat them. The animals in the Amazonian forests gnaw, eat and bury fruits and seeds in exactly the right way to enable them to germinate afterwards. Plants and animals seem to adapt their life perfectly to the needs of their parasites. Whole ecological systems seem perfectly attuned, as if designed to fit together.

Observing the apparent planned nature of ecological systems at any point in time, it is easy to forget that ecological systems developed over the course of hundreds of millions of years. Millions of generations of living beings have lived and died during the development of ecological systems. It is tempting, when looking at any living system now, to interpret its amazingly complex perfection as pre-planned or pre-designed. To interpret it in this way is actually putting the cart before the horse. All ecological systems developed gradually over millions of years.

Any available resource is an opportunity for a living organism. More and more living organisms found resources they could use in order to live and as more organisms developed, the variety of resources increased manifold because life itself creates resources for other life forms. One example of this is on coral reefs, where the coral is the food source for a whole community of animals, such as porcupine fish and parrot fish who clear patches with their grazing and thereby in their turn provide an opportunity for other, fast-growing, coral and algae to establish themselves and grow. A sequence of living species providing resources for each other. Another example is in tropical rainforests, where certain trees and plants depend on animals to gnaw and eat their fruits and nuts so the seeds are released from them, the seeds cannot germinate without this. The advantage to the tree is that the seeds are then spread out over a larger area by the animals. Over time, the evolutionary process developed into the incredibly complex web of life that we now see, and of which we are part.

The highly integrated perfection of living systems can give the impression that this was pre-planned, because we don't see all the species that did not survive and went extinct in the process of forming this habitat; we only see the survivors.

The manner in which life in its habitats develops applies as much to us humans as it does to other species. Human societies and cultures, our agri-culture and technology, all gradually took shape over time. Some traditional religions claimed that our environment was designed to serve us humans. Our current knowledge tells us that this is not the case. Rather, we adapted to our environment, not the other way round. Our evolutionary ancestors discovered more and more ways to use the resources that happen to exist in our environ-ment. All living species developed their ecology like this. Perhaps the reason why our ecology might look more pre-planned compared to that of other species is because we are able to change it to suit our needs more.

Our current understanding of how habitats and their living communities function, indicates that our current way of living will most likely not continue as it is. Therefore, people need to take changes into account when planning for the future. This means promoting changes for the better, and preparing to deal with changes for the worse.

**So one objective of introducing The Natural Religion is to increase awareness of the changing nature of our lives and our environment, and to make sure that we form practical plans that deal effectively with this.** For instance, one direction of change that The Natural Religion strongly supports is to reduce poverty. 25% of our world population live in poverty, that is more than 1.6 billion people. This level of poverty needs to be drastically reduced, and is discussed in detail in Chapter 8. We also need to prepare for such potentially damaging events as climate change resulting in reduced agricultural production and rising sea levels, over-population, and the emergence of new infectious diseases.

We as a species, will in the future most likely continue to change genetically as the world's environment changes. Change of our genetic information via sexual reproduction is, as we can see in the Appendix, comparatively slow. Contrast this with our knowledge information which changes much faster. If the environment which has formed us suddenly changes faster and our genetic evolution can't keep up, then we could go extinct, as the dinosaurs did 65 million years ago.

Time, as pointed out before, is a very important element of evolution. That is why it is so significant that knowledge information can change and adapt so much faster than genetic information (see Appendix). It would seem that our knowledge information will soon have the capability to change parts of our genetic information and makeup. This would speed up our rate of genetic change out of all recognition. For example, research on stem cells, gene therapy, and rejection of transplanted organs has already made advances in this direction.

As we know, knowledge information is a product of genetic information. The genes of an animal allow it to develop the nerve cells and brain tissue so it can gather, store, and use knowledge. Evolution – previously dependent on the relatively slow rate at which genetic information changes – seems to have now found a way to speed up genetic change using its own new invention: knowledge information.

Many people warn against the danger of changing our own genetic makeup. Our genes are so central to our existence that if we make a mistake, or if someone made harmful changes to our genes on purpose, it could have catastrophic results. On the other hand, if we make advantageous changes to our genes this could be a very powerful, and in some instances the only, way to treat diseases. If we could insert genes to treat genetic disabilities or diseases, or use transplant organs from, for instance, genetically modified pigs that would not be rejected by the human body, this would have huge implications for the health and quality of life of very many seriously ill people.

Both the living and the nonliving parts of the Earth's environment will change. We as a species will have to adapt our ecology to these changes or face extinction. Some of these changes could happen at a faster rate than the rate at which our genes are able to adapt. Accordingly, we should not ban human genetic research, but we should monitor and control it so that it is transparent and its power is only used for the benefit of humankind. If human genetic research were to be banned, it would be carried out in secret where it could not be monitored or controlled.

If our earthly environment were to change more rapidly than our genes, one way we could potentially save ourselves as a species would be to speed up our genetic change using our knowledge information. This is in addition to the immediate medical benefits, which having the knowledge to control and change our genes would give us. Could this be unethical in anyway? Not if its effects are beneficial. **So, the conclusion of The Natural Religion is that gene therapy and modification should be applied with great care, but as long as these powerful techniques are open to examination and can be monitored and controlled, their great potential to benefit us needs to continue to be explored.**

## 2.8: Success in life

Life has to deal with the range of circumstances in two ways.

Firstly, each environmental condition has a range of values. Temperature is an example of this. Polar areas have temperatures of less than −50 degrees Celsius (-58 degrees Fahrenheit), and deserts can have temperatures of more than 50 degrees Celsius (122 degrees Fahrenheit). Temperatures can also change over time, be it a day, season or year at any one location on Earth.

Secondly, life, including ourselves, also has to deal with a variety of different conditions. For example, we as humans need food, water, shelter, sleep, safety, and space to live every day of our lives. We also need care, social companionship, learning, personal relationships, emotional attachments to others, a sense of purpose and achievement, mental stimulation and relaxation during the course of our lifetime but not necessarily everyday.

These are just some of the circumstances and factors with which one person has to deal. The study of ecology examines as many of such needs as possible. Ecology looks at how these factors and needs may combine, how they may act contrary to each other, at what stage during an individual's life they occur, which are more important than others, and which are essential and which are optional. All these interacting factors form the life of an individual of any species, including humans. Because there are so many factors some of which may not even be known, there can be a tendency to concentrate on just one or two. This at times leads to an oversimplified understanding of ecology and life. An example of an oversimplified perception of ecology is the widely held misconception of what 'survival of the fittest' means in evolution. The word 'fit' is often taken to mean the strongest, or fastest, or most intelligent. In the context of the ecology and evolution of a species, the 'fittest' are those who have offspring that also produce offspring in their turn. Therefore to be 'fit' in the evolutionary sense is not necessarily the same as the popularly meaning of fit, i.e.: being physically strong, fast or capable great tests of endurance.

Evolutionary 'fitness' means being able to deal with a whole lifetime of circumstances. It means that you survived as a baby, a child, and a teenager. It means that you can find and attract a partner to reproduce with. It means that you give young offspring the love, care, and attention that they need, particularly with mammals like ourselves. Tall strong people are not 'fit' if they are susceptible to diseases. Healthy people are not 'fit' if they don't look after their children properly. Strong and healthy people who give good care to their children will be less 'fit' if they lack learning and education, and poverty reduces 'fitness', no matter who you are. Evolutionary fitness includes all of this and more – it is hugely multidimensional.

At this juncture it is necessary to point out, that gross oversimplifications and dangerous misrepresentations of the term evolutionary 'fitness' have been used as an excuse for disgraceful, immoral, cynical and misinformed policies; such as eugenics, racism, war, cruelty, murder and other crimes against humanity. One such example, is the genocide of Jewish and other minority groups carried out by the Nazis in Germany during the Second World War. The correct meaning of evolutionary fitness was actually turned on its head by the Nazis, and they cynically used this to support their cruel and inhumane dictatorial regime. This is mentioned because these crimes against humanity are at times used to discredit the scientific concept of evolutionary fitness.

**As is clear from the above, the option of The Natural Religion offered here is based on the holistic scientific definition of evolutionary fitness, taking all the different and varying factors that influence evolutionary fitness into account.**

Success of cultures and societies appears to go through cycles. People living in a successful empire often feel themselves far superior to people from other regions, but as time goes by circumstances will always change and people from other nations will gain the upper hand. Throughout history, powerful empires and wealthy nations have come and gone. For example, Mesopotamia in the fertile Golden Crescent, and Egypt along the Nile were centres of cultural, political and military power. Their prominence has long since waned. The Roman Empire, which controlled much of the Mediterranean for more than six centuries, was also superseded. Although China is now on the rise again after centuries of decline, in the 1600s it had grown to be the most technologically advanced country in the world. Its subsequent decline was to see it invaded and part-colonized even into the 20th century. The Maya, Inca and

Aztec empires in America were very powerful and dominated vast regions, but have long since disappeared.

There have been many more rich and powerful empires but all have declined in status for one reason or another. The peoples of these empires were very 'fit' for many centuries, but without exception in all cases circumstances changed and their power and culture declined.

In the present-day world, large amounts of energy are crucial in what are currently the most technologically advanced and economically strong countries. Most of this energy is generated by fossil fuels, but these will run out. Technology for gathering energy directly from sunlight and other renewable sources could be further developed and become economically viable. Many poor and currently economically disadvantaged countries are located in the tropics and have strong sunlight. If sunlight were to become a main source of technological energy these countries could gain great advantages, including possible world dominance, from their strong sunlight. It is changes in circumstance such as these that have caused the rise to power and subsequent decline and eclipse of cultures and nations. For example, agricultural practices in Mesopotamia caused nutrients to be leached from, and salt to accumulate in, surface soils. This caused the decline of the agricultural resources on which these cultures were based, thus contributing to the demise of this historically influential region. Similarly, future shortages of fossil fuels could cause an energy crisis and seriously weaken the position of Europe and North America, today's dominant economic empires.

So, changes occur in the history of human ecology. The most successful and 'fittest' societies or nations at any given time tend not to continue to be the 'fittest'. Other nations then assume that role. The constant change which occurs in general ecological systems, also occurs in human ecology. **For these reasons, the advice of The Natural Religion is that we take into account that success in life is usually temporary, and that we need to expect and prepare for the inevitable changes in our societies and our personal lives in planning the future.**

## 2.9: The dynamic balance of life

Communities of microbes, plants and animals living and interacting together and forming ecological systems are complex and multifaceted. Nevertheless, important environmental factors, such as temperature and humidity, can be identified and described. Most living beings can exist in a range of values of a particular factor. For example, humans can live from subzero temperatures to temperatures in the high 40s Celsius (105–120 Fahrenheit), but most people find temperatures in the 20s Celsius (70–85 Fahrenheit) most comfortable, although depending on whether we are active or not this will vary. Humidity is another factor that influences where plants grow and animals live. Both too dry and too wet are avoided. Most life is found at, or near, an intermediate humidity level, but depending on the time of year and what the particular plant or animal is doing, the optimum value of humidity may change.

This dynamic balance between two extremes, such as between scorching heat and freezing cold and between smothering high humidity and desert-like drought, is one of the main characteristics of the ecology of living species. Life exists in a dynamic balance between two extremes. The extremes are usually very unstable or downright lethal. Both overheating and freezing can kill most living beings. The optimum balance is intermediate and also dynamic because it is rare that it stays static, the optimum changing as circumstances change.

This dynamic or changing balance is also seen throughout our human ecology. For example, individual people can range from being caring to being selfish at different times in their life, or even in one day. In many societies being altruistic is an ideal and being egoistic regarded as abhorrent, but in practice people have to strike a balance between caring for oneself and caring for others. People who are completely altruistic and constantly helping others run the risk of neglecting themselves, this is at the altruistic extreme. Complete altruists run the risk of being used as 'door mats' and taken advantage of to their own detriment. On the other hand, people who are completely self-centred may find that others won't want to help them because helping a selfish person is not appreciated and/or the help reciprocated. Selfish people can be shunned by others and socially isolated, which is ultimately detrimental to them.

In most people's lives, a balance is struck between the two extremes. We dedicate part of our energies and resources to helping other people, but we also need to support our own lives and reserve a healthy proportion of our energies for our own needs. At times, depending on circumstances, this balance shifts one way or the other. A baby cannot look after itself, it is completely dependent and therefore has to be totally egoistic. Also, a person trying to achieve an important and difficult goal, such as studying for an examination, will have little time and energy to help other people. On the other hand, parents with a new baby will shift their balance between helping themselves and helping others in the altruistic direction because their baby needs total care and attention. However, as the baby grows into a child and then an adult, the dynamic balance between the egoism and altruism of the parent will move back to a more evenly balanced situation.

Human ecology operates from the level of our personal lives all the way to the level of entire nations. For example, economics is the resource partitioning of human ecology, in other words it is the way we distribute our resources and it also has dynamic balances between extremes. Neither the extreme of uncontrolled laissez-faire capitalism, nor the other extreme of state-controlled communal or communist systems have proved to be stable. A partially controlled mixed economy, where the control can change depending on the economic climate as described in Chapter 8, have proved to be more reliable and efficient.

Human politics and the power to make decisions and manage a society are also part of our lives and ecology, and also involve dynamic balances between extremes. Outright dictatorships on the one hand, and an extreme form of democracy in which every person is involved in taking every decision on the other, are the two extremes. As discussed in Chapter 9, both these extremes have serious disadvantages and many nations have found that the best way to manage a society is to achieve a balance between centralized control and extreme democracy.

Dynamic balances exist in the ecology of all living beings, be they microbes, plants or animals. As well as the balances in the daily life of living organisms mentioned above, there also exists a dynamic balance within the genetic evolutionary development of whole species. As we know the earth's environment changes all the time, and this also means that whole species have to genetically adapt to these changes. The shifting balance in the genetic evolution of species is best described by an example.

Imagine a bay along the seashore. Half the bottom of this bay is sand and the other half is mud. A type of shellfish lives on the bottom of this bay, filtering the seawater and eating the plankton that it filters out.

Gradually over many generations two types of shellfish develop. One evolves which is better able to deal with the soft mud, and another one evolves which is better able to live on sand. The original species of shellfish was able to live on both a sandy and muddy bottom, but the new types are more efficient and therefore take over from the old type. This trend towards becoming more efficient is common in the evolution of species. It happens because those that are more efficient produce more successful offspring than their less efficient contemporaries, so they increase in the population as a whole and they eventually take over. This efficiency is usually achieved by being very specialized and competitive in a particular manner of feeding and way of life, but losing the ability to compete in other conditions.

Returning to our example, one of the two new species specializes in muddy conditions and the other one becomes a specialist in sandy conditions. The original species which is a generalist, now only survives on the edges where there is a mix of sand and mud. Being more efficient the specialist sandy and muddy bottom shellfish now live in higher densities compared to the number of the original generalist shellfish. The population of the original generalist shellfish was at a lower level to start with because it was less efficient, but now it is reduced even further because the sand and mud mixed areas only exist around the edges.

Now suddenly the environment in the bay changes because torrential rains cause massive landslides inland from the bay. The landslide cascades down a hillside and into a river. The mixture of sand, mud and water forms a torrent that thunders downriver and enters the bay in a seething boiling flood. As the river water races into the bay it spreads out and then slows down. This has the effect of allowing a sand and mud mix to fall to the bottom and settle over the sand as well as the mud. The whole bottom of the bay is now covered in a mix of sand and mud.

Because the shellfish species that was specialized for sandy conditions did not need to be able to live in mud it had lost the ability to do so, and now it is not able to deal with the sand and mud mix which the river deposited in the bay. The same has happened to the species specialized for the mud, it is not able to deal with the sand mud mix either because it has lost the ability to deal with sand. Now the generalist species all of a sudden comes into its own again, because it is able to live in a mud and sand mix and so it once more takes over the whole bay. Being a generalist, its population in the bay will not be as high as the sand and mud specialist populations that used to live there, but the generalist still has the ability to deal with both mud and sand.

The evolution process does not stop there. As generations of generalist shellfish come and go, more efficient specialist species will develop again. The bay because of the mud and sand landslide, has now become very shallow around the mouth of the river. The other half of the bay is still deep with a steep drop off between the shallow and the deep. The two types of shellfish that now evolve specialize in living in shallow water and in deep water. The old generalist shellfish just about survives on the steep incline between the shallow and the deep.

Once more, circumstances change because a hurricane sweeps into the bay sending huge waves ripping into the shallows and cause the settled mud and sand to be spread all over the bay. Now most of the bay is again intermediate in depth and again the shallow and deep specialists cannot genetically adapt fast enough to this rapid change in water depth, and the old generalist can once more out-compete its two specialist rivals.

So, in evolution, a dynamic balance swings over and back between the two extremes of, at one end, generalists with high adaptability but reduced efficiency, and, at the other end, specialists with high efficiency but reduced adaptability. The way species evolve is therefore also dynamic, with the position of balance at any one time dependent on how fast and how often the environment changes.

The changing nature of evolutionary adaptedness and 'fitness', and how this applies to human beings is often not fully appreciated. This lack of understanding is not just of theoretical or academic relevance, for instance it contributed to the views of the already mentioned eugenics and Nazi movements which resulted in much human suffering and death. It also continues to contribute to racist attitudes today. It is the aim of this book to clearly explain how and why ecology and evolution function the way they do, so that a lack of understanding doesn't contribute to more tragedy. The example of the generalist and specialist shellfish, and the dynamic balances of our human economic and political life, show how life operates. What is successful now will have to change in the future in order to survive, and today's underdog can be tomorrow's winner.

In contrast to traditional religions, which often explain human life as being influenced by good and bad forces struggling against each other, **the practical advice proposed as part of The Natural Religion is that realistically in many aspects of our private lives as well as society at large, we need to identify a balance between two extremes. This balance should not be static but needs to be able to move towards one extreme and back again towards the other when circumstances demand it.** The balance of these aspects need sensitive adjustment, many of the decisions we need to take in life are about judging where a particular balance should lie. While it is impossible for The Natural Religion to identify exactly where this balance lies in each instance, it can identify extremes.

In the next chapter we concentrate on the fact that our individual lives begin and end, and the influence death has on our emotions.

# Before and After Life

*⚜* Death is a fundamental part of Earthly life *⚜*

## 3.1: Taboos about death

Death is when we stop living. We also don't exist before we start living. So, we don't exist both before we start living as well as after we stop living. Death is contrary to most people's basic instinct to live and stay alive and therefore in many cultures people fear death. We have an inbuilt urge for self-preservation, yet our self-awareness tells us we will die. This creates a conflict of feelings and emotions. To delve into death and discuss it in detail is at times seen as negative and defeatist, even macabre and unhealthy. In general conversation, death is often only referred to in passing or obliquely because people are apprehensive about it and do not find it a comfortable topic for conversation. As a result death is not often discussed in detail or fully understood.

Yet as we will now see, death is as important to our existence as birth is.

Religions traditionally try to resolve our anxiety about death. Supernatural beliefs are often used to help with this. Numerous different variations of beliefs exist about what happens when one is dead. As mentioned in Chapter 1, many of these beliefs contradict each other and we cannot check which one is correct. Beliefs vary in terms of: in how many human bodies one spirit or supernatural entity can reside over time; where a spirit goes after the person has stopped living; how long the spirit survives; whether souls or spirits are also found in other animals and plants. Fear and apprehension of death have been used in the past by religions as a way to gain influence over people.

Various religions state that there is a supernatural afterlife when you are dead and that either good things or bad things can happen to you in that afterlife, and furthermore only they can help you to avoid those bad events after you die. If a person believes this, it creates a great dependency on, and gives power to, those who manage and administer that religion.

An argument that is used to encourage believing in the existence of a supernatural afterlife, is that it persuades people to be unselfish and live good lives if they believe that they will be rewarded for this after they die. The trouble is that there is no evidence that an afterlife exists and increasingly people don't take this belief seriously. **What is certain is that if we live good lives and help others, our actions can benefit our own lives as well as the lives of others. The meaning of our lives is created by our own actions and the legacy we leave behind.**

We know for certain that we are now alive and also that we will stop living. Our awareness of our mortality and the anticipation of death can cause us to feel sad and fearful. We feel grief and deep loss when a loved one dies, and we also fear our own death.

The Natural Religion, like other religions, also attempts to help us deal with death. This chapter looks at our knowledge of death and the deep instincts and emotions it evokes in us. It approaches death from a realistic point of view, discussing the psychological effects that death has on us and the central role that it plays in our biological system. It attempts to help people to deal with death through an open and honest discussion of our feelings about death and why it exists. The ideas expressed in this book take an agnostic point of view towards supernatural concepts. **The psychological support and the meaning of our individual human lives offered here do not include either a utopian or a tormented afterlife. It is based on knowledge and reasoning leading to realism and greater dependability, thereby reducing apprehension and anxiety about death and so also contributing to our quality of life.**

## 3.2: **When a loved one stops living**

We humans are a social species. We form close emotional bonds with the people with whom we share our lives, as many other animal species do. We mostly live together and do things in cooperation with each other. Friendship and companionship particularly of families and friends are crucial to people's general happiness and emotional well-being. When people with whom we have a loving or close relationship stop living, their love and companionship is suddenly taken away. The emotional bond of the relationship is broken. Our brain gives us the realization that they will never come back to us and causes deep mental anguish and emotional distress. At times it has even been described as causing a physical pain. Sometimes it is the grief experienced when a relationship is broken by death, that makes the surviving person fully realize the full depth and importance of the relationship.

The passage of time has the effect of lessening, and eventually removing the emotional upset of grief. But the memory will remain. The period of time for grief to fade depends very much on the individual concerned, as well as on the quality of the relationship that is broken. Estimates for grieving periods vary from three to seven years and sometimes even nine or more years.

Humans are not the only animals that experience mental distress at the death of an individual well known to them. In fact, distress at the end of close relationships are known for a wide range of animals. Examples are reported from: elephants, peregrine falcons, dogs, horses, dolphins, lions, rhesus monkeys, beavers, gorillas and chimpanzees. In some of these cases the distress about the loss of a family member, mate or companion even ended in the death of the animal showing the distress. Whether other animals feel grief in exactly the same way as we do, we cannot be sure. We can't communicate with other animals the way we can with other people. However, observers of these reported examples were in no doubt that these animals were experiencing emotional distress and simply the absence of an individual caused this. We don't know if other animals understand death and that life for an individual exists for a period of time only, or if they do whether it is in the same way as we understand it. Close relationships do, however, seem to exist in a wide range of other animals besides ourselves.

The human brain has developed to a high degree of sensitivity. It is this same sensitivity that allows us to be aware of death and is possibly the reason for our prolonged period of grieving. Other animals can be distressed when they lose a companion, but they may not be aware of death as humans are. In all likelihood, we understand what death is much better than other animals. For example, we humans can understand the difference between the mere absence of a person and death. However, while we grieve for people with whom we have a relationship, we tend not to grieve about all our long-dead ancestors, because we did not know them personally.

Ancient prehistoric burial sites have been uncovered in which food, jewellery, flowers and other items were found in the grave together with the person who was buried there. Some speculate that maybe such an elaborate burial might have been part of a religion or belief in an afterlife and the supernatural. What is more certain is that the trouble to which these people went when they buried the deceased does point towards an emotional bond with that person. Whether these ancient ancestors actually had beliefs in a spiritual existence after death is not known, but what these prehistoric burials do suggest is that people grieved for those whose lives had come to an end. A burial, cremation or other death ceremony marks the end of a life and offers an opportunity to show appreciation and gratitude for the life of a person and recognition for that person's achievements. These ceremonies go some small way towards helping and supporting grieving family and friends.

To recall and to be thankful for the achievements of people who are now dead is a fitting way to remember them because these achievements often continue to influence ourselves and future generations and give meaning to people's lives long after they have stopped living. For example, agricultural techniques, general infrastructure and technology of all sorts, as well as ideas and inventions, are given by one generation to the next. Our human knowledge has accumulated as a result of the achievements of many generations of our predecessors. Much knowledge from which we now benefit came from people we never knew.

Every person we know during our lifetime influences us to a lesser or greater extent. Their ideas, attitudes and personalities all influence us. Those we know well and with whom we have close relationships influence us more than those we don't know that well, but every person we meet becomes part of our experience of life. When a loved one stops living, we lose their company, but their influence on our attitudes and ways of thinking, and the benefit from their achievements, continues with us after their death. In this way death is not a cut-off or an end. These memories and influences that continue to be with us after their death are passed on by us even further to others. Ideas and ways of doing things continue from one person to the next for many generations. The influence of a person, and the meaning of their life, can live on in the culture of a family, local society or a nation long after they have stopped living.

One simple example of how ways of thinking and doing things are transmitted through many generations in a culture is the inflections, accents and dialects of a language, as well as the language itself. Another example is technology, which is a physical achievement of people who have long since died and which we continue to use. We don't know who invented the first wheel or steel wood-saw; all we know is approximately when and where they lived, but their lives still have influence and meaning today.

Most people will experience bereavement during their lives. Only those who stop living at a very young age, or who are very lucky, will not experience the death of a loved one. Death has always been part of human life. The death of a loved one and the grief that comes with it was, of course, also experienced by our evolutionary ancestors. It is assumed, therefore, that we have evolved some emotional mechanism to cope with the end of a close relationship. Yet, our grieving period can be long. Maybe this shows that the importance of relationships has also evolved, with grieving becoming worse as relationships became stronger. Nevertheless, our ancestors did have the strength to endure grief while the passage of time weakened its severity, like the majority of people today. It should be remembered that close, loyal and loving relationships help us and give us great support. We can achieve more with the help of these relationships than without them and they are of great benefit to us. But when the relationship bond is broken, it causes grief, particularly if the relationship was an emotionally close one.

**Grief is an intrinsic part of relationships. A relationship that is so important and good for two people when both are living also causes much grief when one of these people dies. Death is part of life.** Before we were born, we did not exist and we again don't exist after we stop living. This is the way our system of life works and we humans may be the only form of life that is fully aware of this. The Natural Religion, like other religions, cannot take away grief. Grief is a very distressing consequence of what are otherwise very good aspects of our lives: our close personal relationships. Our lives do not continue forever, and both grief and death are intrinsic parts of life. Both are as natural as life itself.

**The viewpoint on which the concept of The Natural Religion is based is that the function and purpose of life is to live. The reason we have death in the first place is because life exists. Therefore when one person stops living, the rest of life should continue living.** While we are alive living is what we have to do and should do. When we are grieving, time will gradually weaken the severity of our grief and it will eventually fade away as we continue to live life.

Although the end of a life seems very final and irrevocable, the ideas, attitudes, achievements and influences of those we have known will continue on with us after they have stopped living. **Our loved ones continue to be of importance in our lives and continue to help us with their legacy of influences long after they are gone. Our recognition of the existence and importance of these legacies is based on observable and verifiable knowledge. We don't have to rely on beliefs held on faith to realize and recognize the continuing importance of a person's life after they stop living.**

Death ceremonies or funerals in most religions have supernatural beliefs as part of them. However, the alternative religious options set out here are not supernatural in any way, but funerals, whether supernatural beliefs are part of them or not, are very important for those who are grieving. To mark the passing of a life is just as important to an agnostic as to a believer or atheist, and the importance of this is not reduced for someone who attends a funeral of a religion with supernatural beliefs they believe in. Respect for and appreciation of a person and their past life do not depend on beliefs and can be expressed whether a belief religion is part of the funeral or not.

Although no religion can take away a person's grief, the following suggestions are offered to try to help those who are grieving:

- Focus on the day-to-day practicalities of living life.
- Rely on our inbuilt coping mechanisms for grief.
- Do not forget that much of what the deceased person gave to us during our relationship will continue to be with us.
- Appreciate these memories and influences that are a great emotional and practical gift.
- Be grateful for the relationship we had with our loved one.
- Think of death as a normal and natural part of life.
- And remember that, while time will gradually take away the grief, it will not take away the memories.

## 3.3: **When we ourselves stop living**

We start living first as a fertilized egg, then we develop into a foetus, after which we are born. Before we start living, we are not alive and also after we stop living we again are not alive. Our life lasts for a certain period of time only. Since we are self-aware, we know that we will stop living. Most people do not want to die because humans, like other animals, have a strong urge for self-preservation. We are alive because our ancestors also had this instinct for self-preservation, so it is normal for people to have an apprehension, even a fear, of death. To preserve our own life is ingrained in us – it is us. To stop living is usually seen as the worst thing that can happen. We do know, however, that we will definitely stop living at some time in the future. Yet most people strive to keep alive for as long as possible. Time is an essential factor in the whole living system, including in our own lives. **Once we are alive what we have is time. No individual is absolutely sure how much time they have, whether it be minutes, hours, days or years. Life happens over time.**

We don't worry about the time before we started living, but we do worry about death after we stop living. When we try to remember back to before we were alive, we don't have memories of suffering or distress. In fact we don't have any memories at all.

A number of religions say that after we stop living, there is a kind of continued spiritual existence or afterlife. The idea of an afterlife can be reassuring because the person who believes in it, does not see the end of living as a final end. It satisfies the urge for self-preservation in us. However, some religions claim that not all of the afterlife is good and that if one does not live a good life then the afterlife can be full of suffering. This again introduces a fear of death because few people have never made mistakes and lived a completely 'good' life.

As mentioned before, it is not possible to know if an afterlife exists or not. However, some feel that our dreams are an indication that we have a spirit or soul. Dreams are a phenomenon of life. Our brain gives us dreams during our sleep. Humans are not the only animals that have dreams. Dogs and cats make movements in their sleep, even make noises, just as people mumble words in their sleep sometimes. Electrical monitoring machines have recorded particular types of electrical activity and stages of sleep in the human brain at the time that the person being monitored reported that they were dreaming. Similar electrical activity and stages of sleep were recorded in the brains of a variety of higher animals, such as cats, rats and sheep, as well as showing outward signs of dreaming as most dog owners will know and also including other animals such as elephants. From this we can conclude that, dreams are physical events that occur in the brain during sleep. Dreams are therefore no indication of a spirit or a soul in us or any other animals.

The feelings we have when we think about our own lives and the fact that we will stop living cause deep emotions. The meaning and relevance of our lives are amongst the most important things we think about:

- What am I living for?
- Does it matter what I do with my life?
- If I had not lived, would it have made any difference?
- Why am I me? Could I have been someone else?
- Can I accept losing the life I have?
- What will the end of life be like?
- What is death like?

These are all serious questions that are of deep importance to us because they are about the deepest roots of our existence. They are at the heart of our being.

First and foremost, we are alive. We have time during which we live. We, as mammalian primate animals, have evolved in such a way as to use our time carrying out the activities of life. **Life is not about living as short a time as possible and doing as little as possible: life is all about living as long as possible and doing as much as possible.**

One of the most important things that our ancestors did with their lives was to live long enough to reproduce successfully. They also achieved enough in terms of accumulating resources to enable their offspring to reproduce. The fact that we are alive proves that. Those of our ancestors' offspring who inherited this instinct also produced offspring, as we in our turn also inherited the drive to lead an active life and reproduce. Prolonging life and having expectations and actively pursuing them is a strategy that has been evolutionary successful. This is the reason why we are alive. This is also the reason why we are physically and emotionally designed to try to live a long life and actively pursue achievements.

So we have inherited certain traits that proved successful in the past. Attempting to achieve goals is one of them. Mentally, we are programmed to try to achieve things. Achieving our own personal goals makes us happy and contented. Helping our close families gives us great satisfaction, and being able to make a significant contribution to society gives us a feeling of pride and self-esteem.

Another aspect of achievements is that they continue to be of benefit to people, long after the person who achieved them has stopped living. At all times in human history, people have received ideas, attitudes, methods, tools, technology, and knowledge from their predecessors. And, of course, every living person is the product of the reproductive achievement of his or her parents and ancestors. In this way, a person's life continues to have significance long after they have stopped living. If they had not lived, these achievements would not have been made or would have been made in a different way.

To the question: Could we have been someone else? The answer is 'yes'. We could have been someone else, extremely easily in fact.

Each individual person receives a particular set of genes from each of their parents when the sperm and egg combine to start their life. During sexual reproduction, different combinations of chromosomes can be selected. The genes also change position on the chromosomes and can line up to form different combinations when the mother's egg and the father's sperm is formed. As a result, many multiples of thousands of combinations of the mother's and the father's genes are possible. One of the thousands of possible combinations from the mother's side combines with one of the thousands of combinations from the father's side to form that particular fertilized egg. This gives a huge potential of genetically distinct fertilized eggs. Each of these would be a different person, but it is, of course, completely impossible for human parents to actually have all these children. Out of the hundreds of thousands of possibilities, that particular person is the one that will develop and live because of that particular combination of genes. So each human has been the lucky one who received life. **For every living person, many multiples of thousands of other possible humans did not get to live. The fact that we are alive means we are the genetic lottery winners. We are the lucky ones.**

We have a fear of losing our life. It is our instinct for self-preservation. However, the fear of losing one's life is different from the fear of being dead. We are designed to want to stay alive and to prolong life. To accept that our life will stop goes against one of our most basic instincts. However, there are two things to consider about the fact that our lives will stop.

The first is that we can prepare for death to a certain extent. If we are happy with the way we lived our life, with the family of which we are part, with the children we raised, with our achievements, with the type of experiences we had, these all satisfy our inbuilt urge to do things with our life. Every person is not able to fully achieve each ideal during their lifetime, but if we try and fight against any misfortune and maybe only partially attain our goals, then we will have lived the life we were designed to live. If we feel that we have used our efforts to the best effect that circumstances allowed, we will feel that we have made an impact and our lives had meaning and purpose. This approach to life made our ancestors successful and therefore this has been passed on to us.

By trying to make sure that we can be content with our achievements and not be sorry about the way we lived our life, we can prepare to some extent for death. We will be able to accept losing life better if we feel we made as good use of the time that we lived as possible. The achievements of our predecessors gave rise to and supported us in our life. So also our achievements will give rise to and support the people who live after us. The fulfilment that past achievements give us goes some way to offset our apprehension of death and inbuilt urge to survive. Part of us and our efforts will continue to exist when our children and/or our achievements and influences continue on after we stop living.

The second thing we should remember about dying is that death is a completely natural and integral part of human life. In fact, death is an essential part of every individual of every species that ever lived on this planet. The fact that individual living organisms don't continue to live allows species to evolve and adapt to the changing Earthly environment. Helped by death, life has developed and changed and accumulated its store of genetic information from one generation to the next. For example, life would very likely not have developed into having a brain that is self-aware had it not been for death. Every living human being gains great benefit from death, including the most basic one of all: the fact that we exist.

Paradoxically, our self-awareness, which death has helped us to evolve is also the cause of our apprehension at the prospect of our death. Dying is nevertheless the natural culmination of our life. It is the normal conclusion to the time we are alive. There is nothing abnormal about a natural death; we just have an inbuilt drive to prolong our life.

In humans, our body starts to decline in the latter part of our lives. This development of frailty is a natural physical process that our bodies are genetically programmed to carry out. Gradually we become less vigorous and draw back from active life around us. As we grow old, we also change our role in our family and society. On the whole, older people's contribution changes from active participation to more passive assistance. The knowledge and physical resources that older people have accumulated during their lives, are often given to the younger generation to assist them to rear their children and generally help them in life. In families, older women tend to actively help to look after the young children for longer than the older men. But gradually, all older people draw back from active life, following an inbuilt progression by which we move towards stopping with living.

The fact that our bodies and minds are designed to decline is an indication of how important it is that we stop with living after a full and active life. Natural death is not a negative whereby life just wears out; it is as essential a part of the whole biological system as birth. Death is a basic part of evolution and, as we know, we need evolution because the Earth's environment changes all the time. As individual human beings we contribute to our family, our society, our species and to the system of life in general by stopping with living. The fact that we stop with living is a positive contribution to life on Earth, not a negative failure. We benefit from the fact that those who went before us stopped living, just as those who come after us depend on us to stop living.

This approach to death is based on what we know, rather than on suppositional beliefs. We know we affect the life of all the people who know us and also many who do not know us. Our effect is carried on in many ways, consciously as well as in ways that we may not even realize. For example:

- The ideas and feelings we give or instil in others.
- The memories we leave behind.
- The knowledge we share with others.

- The inventions we create.
- The resources we bequeath.
- The next generation of babies, children and adults we produce.

All these achievements give our lives significance and our existence meaning long after we have stopped living. In fact, our current human ecology with all its technology and great impact on our planet, was created by the combined efforts and achievements of many people's lifetimes. **The way we live now was created by our predecessors; that is the meaning and significance of their lives. So also it is the meaning and significance of our own lives that we prepare our world for future generations.** All these achievements are real. They can be examined, checked and verified. These are the real results of people's lives. This is why The Natural Religion points to achievements as a reason for living and a meaning of life.

People need not fear the end of living. To accept that one's life will end is not the same as losing one's drive for self-preservation or the will to live. As mentioned before, when we die a natural death, we are playing our part in the system of life on Earth as much as when we were born. We are participating in the future life of our families and communities by stopping living after a normal lifetime, just as our predecessors continue to influence our lives. **Living-time is what we have. To be alive is very special and we can live with determination and gusto and enjoy life as much as our situation will allow. To try and achieve our expectations and strive for contentment during our lives is not incompatible with accepting that our lives will end. After all, death is part of life. It is because of life that death exists, but the reason for our existence is to live.**

**Part of the set of ideas collectively called The Natural Religion, is that both before, as well as after our life we don't exist. There are no consistent and verifiable indications that not being alive is anything more than just not existing. Therefore, the conclusion arrived at here is that since we simply don't exist before and after our life, life is not a preparation for a supernatural existence after death. There are no realistic indications that our life is a test for events that will happen to us after we stop living. What we do with our lives affects us personally only during our lives.**

The way some religions claim we continue in an afterlife is by means of a supernatural spirit or soul. As mentioned in Chapter 1, the existence of such supernatural entities are not verifiable and religions also differ and contradict each other about the various types of afterlife, so it is not possible to know which, if any, of the beliefs held on faith regarding one's own death is correct.

The way The Natural Religion strives to reach conclusions that can be taken seriously is, as we know, to base them on knowledge and reasoning. **One advantage of viewing our life as our only existence and not as a preparation for another supernatural existence, is that it can help us to fully focus on the time we have alive and to plan it and live it constructively and actively. This view sees our life as the main event, not as a dress rehearsal.** We should enjoy life as much as circumstances will allow, particularly by attempting to realize goals including ones that benefit others.

We have knowledge about our lives because we tell each other about it. However, after we stop living we cannot give knowledge information about being dead because our senses and our brain have stopped working. We therefore cannot say what being dead is like. However, we also did not exist before we started to live. Therefore, to try and judge what death will be like after our lives, we could try to remember what it was like before we lived. But we don't remember anything from before we were born, nor do we remember anything from the time before fertilization of the egg from which we developed. This is the only knowledge that we have about what death could be like. In the absence of any other verifiable information it is not unreasonable to assume, that after we stop living will be the same as before we started living. So we can conclude that we will not experience being dead. We probably won't feel anything. We will just not exist and therefore there is no need to fear being dead.

We cannot say what it feels like at the point when life stops, for the same reason that we don't know what being dead is like. People have reported experiences what are called 'near-death experiences'. However, these, by definition, are reported by people who did not die, so it is not clear whether feelings at the point of real death are the same. If we think of the feeling that we have when we fall asleep, we cannot remember the exact point at which we fall asleep or what it felt like. The point at which we die may be similar to the point at which we fall asleep, in other words, we just don't feel anything.

Unfortunately, it is possible to suffer while we are alive. We know suffering is part of life, but as we don't experience anything when we are not alive we don't suffer either. **It is presumed that our experience of being dead is the same as that before we lived, and also that when we die it is the same as falling asleep. In both cases we simply don't feel or realize anything.**

We have an instinct for self-preservation and try and prolong our lives, and therefore we can fear death. One way to help overcome this fear, is to live our life in such a way that at the end of a normal lifetime, we can look back on our life with satisfaction and contentment about what we have achieved and this will help us to accept the fact that we will stop living.

**The approach of The Natural Religion towards our own death, therefore, is that it should not be feared. Before our life began, there was no suffering and there are no indications that we suffer in any way after we stop living. Not fearing death makes our life happier and acceptance of the end of our lives easier, thus allowing us to live our life to its full potential, and without fear or apprehension.**

## 3.4: Implications of the end of life

All living beings that ever lived on Earth have stopped living, and so will those that are alive today including ourselves. This knowledge of our own death and that of loved ones is difficult to deal with, but death is such an integral and essential part of our system of life that we gain a better understanding and appreciation of life if we understand why we are genetically programmed to stop living.

As mentioned in Chapter 2, life developed the process 'evolution by natural selection' in order to keep up with the changes in our Earth's environment. In sexual reproduction, each offspring is slightly different from its parents. The environment then favours those offspring that are best adapted to circumstances at that time and the whole population changes in the direction of better adaptations.

If genetically programmed death did not exist, the only life on earth would probably be something similar to the first-ever life forms. These primitive forms of life would probably die out whenever the environment changed beyond what they could deal with. Then new primitive life forms would most likely develop again from non-living materials, only to go extinct again as the Earth continued to change.

Life of this type may, in fact, have existed, but did not survive the Earth's changes. Just one type of life – that which can change through reproduction and evolution and death – has survived Earth's unpredictable environment. Life that does not change will go extinct and the next form of life will have to begin from the very start all over again. Life that changes at each generation is able to build up its genetic information and improve on previous generations. We humans, and the other living species with which we currently share this planet, owe our existence to this process of genetic development. Even at the developed stage of life which we are at now, we need death. If we did not have death, populations would become so large that eventually we would go extinct due to lack of food, space and other resources.

A population in which individuals stop living after reproducing adapts and become efficient faster and out-competes another population that adapts more slowly because older individuals stay alive. The basic reason why individuals need to stop living is because the Earth's environment changes and therefore life also has to change or it will go extinct.

The set of ideas about our human existence put forward in this book and named The Natural Religion, are primarily about life and how we live it. As death is an integral part of the way life developed, and because we are genetically programmed to stop living, this chapter discusses death.

The ideas outlined in this chapter show that life needs individuals to stop living at the end of a normal lifetime in order for life in general to survive. **We all stop living as part of our role in the life of our human species and in the biological system in general. We need death at the end of our lives just as much as we need birth at the beginning of our lives.** We should, at the same time, appreciate how truly amazing life is and not let death overshadow our lives. The meaning of our lives is what we can achieve and we can approach it positively by making as much use of the time we are alive as we can. Death is not a negative failure of life, instead it is essential in helping life to adapt so it can survive Earth's changing environment. We will now look at how our ancestors adapted and survived.

# Chapter 4

# Our Origins

🐾 Human evolution is the history of our brain 🐾

## 4.1: What are we, and why?

We are part of life on Earth and as discussed in the previous two chapters, fundamentally our life works in the same way as that of all other living beings on Earth. From what we know of life, we have a common ancestry with all other forms of life on planet Earth. The more closely related we are to a particular living species, the more recently our common ancestors with them lived. Amongst the questions that religions address is: where did we come from, and how did we come to be what we are? We now have the knowledge to be able to answer these questions without having to use supernatural guesses. This chapter discusses the knowledge we now have about our human origins.

We are part of the Animal Kingdom. Within the Animal Kingdom the species of animals are classified according to how closely related they are judged to be. We humans have our place in this classification which is as follows:

Kingdom: Animalia
Phylum: Chordata
Subphylum: Vertebrata
Class: Mammalia
Subclass: Eutheria

Order: Primates
Suborder: Anthropoidea
Family: Hominidae
Genus: Homo
Species: *Homo sapiens sapiens*

This classification to some extent also gives an idea of the stages which our ancestors went through before we evolved as a species. We evolved from animals with a backbone, which gave birth to live young and suckled them on milk. As our evolutionary ancestors developed, their body shape and functions became more and more like we are today. We have stereoscopic eyes, our hands can grasp things and our brain increased in size as our ancestors evolved from one species to the next.

As with all other living species on Earth, our bodies and capabilities are the result of the environment in which our ancestors lived. The circumstances of our ancestors' lives dictated how they found their food, how they found shelter and how they dealt with the diseases with which they had to contend. As conditions in their habitat changed, so also did our ancestors. We are the descendants of those who lived to produce offspring. Their bodies and minds were capable of feeding and protecting their family in the environment in which they lived. **The environment of our ancestors is reflected in our bodies; our bodies hold the clues to what that environment was like.**

## 4.2: Our ancestors

Our knowledge of what our ancestors looked like, comes mostly from fossils. Fossils are formed when plants or animals are buried and chemicals from the ground seep into the plant or animal remains. These chemicals gradually replace the remains and they then harden in that shape. Those parts of living organisms which tend to survive longest underground, like bones or shells, are most likely to have time to fossilize before they rot away. It is important for the interpretation of the sequence of our evolutionary ancestors in time, that we make an estimate of the age of the fossil. There are many different methods to estimate the age of the particular rocks and soil in which a fossil is found, as well as the fossil itself. Some of these involve gases from the atmosphere or chemicals from volcanic eruptions which change radioactively from the time that they were trapped in rock, fossils or ice. The rate of these changes are known, so the time since these chemicals were trapped can be measured. We also have information from biochemical analyses about our evolutionary ancestors. Certain proteins change and mutate at a steady rate. One of these is the DNA of the chromosomes in the mitochondria of a cell. Mitochondria are little organelles inside a cell which look after the energy supply of that cell. The mitochondria and the nucleus and other organelles and structures together make a cell work, but the mitochondria are not in the nucleus, so mitochondrial DNA is not part of the DNA which makes up the chromosomes in the nucleus.

The DNA in the nucleus of a cell changes every generation during sexual reproduction when a half set of chromosomes in the father's sperm joins with a half set of chromosomes in the mother's egg. The new offspring only gets mitochondria from its mother in the egg. The father's sperm is much smaller than the egg and only has chromosomes for the nucleus of the egg in it, it does not contribute any mitochondria. Because mitochondria only come from the mother they don't change during sexual reproduction, but they do change over time at a gradual but steady rate. So by comparing the differences between the mitochondrial DNA of a range of people from all over the world, we can get an idea how closely they are related and this gives us an indication where their ancestors came from.

Our interpretation of fossils, together with the biochemical findings, give us some indications as to who our ancestors were, and where they lived. Estimations are made by taking all the different pieces of information about the fossil into account, such as: the exact location where it was found, its shape, and its estimated age. The evidence on which we base our ideas as regards our evolutionary origins may not seem very strong when we look at one part of a fossil skull, but once we compare it to other fossils and take the biochemical evidence and ageing into account as well, a clearer picture emerges of what happened in our past.

We cannot travel back in time, so the only indications we have about our evolutionary ancestors is from these pieces of physical evidence. Many ideas have been suggested about where we humans come from. Our current evidence indicates that our primate ancestors from 50 million years ago, were like our present-day lemurs. Today many different species of lemurs live on Madagascar and neighbouring islands off the east coast of Africa. Our lemur-like ancestors developed into monkey species, after which they developed into ape-like animals by approximately 20 million years ago. These lemur-type ancestors of both present-day monkeys, apes and ourselves most likely lived in forests, they don't appear to have walked erect. This is important as it gives us an indication that the life they led up to this stage seems to have been in forests. They used their hands and feet to climb and had stereoscopic vision to judge distance from branch to branch.

After our ape-like ancestors, the next series of ancestral fossils which were discovered are much younger. They are estimated to be between 4 million and 1 million years old. From the shape of the skull and the various parts of the skeletons which were discovered it seems that these ancestors walked upright. Fossils of this type were all found in Africa, and are generally classed as the first human-like species. The upright stance is taken to indicate a change in habitat and life style. The oldest of these new human-like types has been named *Australopithecus afarensis* and lived as long as 4 million years ago. The famous partial skeleton found in Ethiopia in 1974, which was given the name Lucy, is deemed to be a member of this species.

Other fossils from this same period are thought to belong to different species. The australopithecine species that is generally accepted to have been the most important is *Australopithecus africanus*, the fossils of which are thought to be between 4 and 1.5 million years old. Another species type is slightly younger, with fossil ages estimated to be between 3 and 1 million years old. This species has been named *Australopithecus robustus*. It would seem that for approximately 1.5 million years both these ancestors were alive at the same time. *Australopithecus robustus* was more heavily built and, judging from its teeth, ate plants and fruits rather than meat. *Australopithecus africanus* on the other hand was lighter in build and probably ate meat as well as plant food.

More australopithecine species most likely existed. The fossils are not all the same and opinions vary about whether certain differences between them should be interpreted as belonging to separate species or to the same species. Either way, as species develop, linking stages have to exist and their characteristics will be intermediate between two species. In fact, we can only designate species on the basis of the particular fossils which we happen to have found. Maybe if other fossils were found we would have designated different species.

What we do know is that in Africa a group of species, the australopithecines, existed which walked upright. This means that they had their hands free to carry things and to handle tools. They appear to have had tools made of flint and bone, so they must have been able to make these. Many ate meat as well as plants, berries and fruits while others were most likely vegetarians and did not eat meat. Australopithecine teeth were different from their ape-like ancestors, but also from our more recent ancestors, which supports the idea that they were an intermediate stage between these two. Like many other animals, including ourselves, they probably had social groups and may have looked after each other's children.

What is interesting is the extent to which the brain had developed in the australopithecines compared to ourselves. The australopithecines had become much more human-like compared to the monkeys and apes from which they most likely developed. Their upright stance in particular would have made them more like us, but australopithecine fossils indicate that their brain although a bit bigger than that of apes had not yet started its big human-like increase in size relative to the rest of the body. The australopithecines were therefore presumably more advanced in using knowledge information than other primates of their time, but compared to ourselves, their brain was still quite small. In the animal kingdom our own brain capabilities are extraordinary even compared to those of our closest living relatives, the chimpanzees.

What the change from monkey and ape-like primates to the australopithecines indicates is that the australopithecines occupied a new habitat in Africa. A habitat in which climbing had become less important and walking became more important. To be able to walk unhindered, and at the same time carry objects, was an important advantage for our ancestors. Our own ecology is hugely dependent on technology and use of tools, the vast majority of which we utilize with our hands. **If we had a big brain but no way to translate our thoughts into real physical advantages, the potential of a big brain could not be realized. So it is interesting that our fossil finds indicate that walking upright and the freedom to use our hands came first, and only then did the biggest evolutionary development of our brain take place.** Knowledge needs to be converted into real physical resources before it is of any advantage, and we mostly do that with our hands.

Another group of fossils have been found with characteristics which are much more like humans than the australopithecines, particularly in the skull. These have been designated as being part of the same genus as ourselves, the genus *Homo*. The oldest of these fossils is thought to have belonged to a species now named *Homo habilis*. The fossils of *Homo habilis* are estimated to be between 2.5 and a little over 1.5 million years old and have all been found in Africa. This means that *Homo habilis* lived at the same time as some of the Australopithecines.

The difference between *Homo habilis* and the Australopithecines is in the larger size of the brain and possibly also a somewhat thinner and taller body. Not everyone who has studied these fossils agrees that *Homo habilis* should be in the genus Homo. Some feel that *Homo habilis* should be included with the Australopithecines. However, whether *Homo habilis* should be in the genus *Homo* or in the genus *Australopithecus*, it still shows a progression towards a greater brain size and a more upright thinner body as time went by.

The next group of fossils which have been given a name belong to the species *Homo erectus*. The oldest of these fossils are estimated to be just over 1.5 million and the youngest between 400,000 and 200,000 years old. *Homo erectus* is the first hominid which seems to have migrated out of Africa and its fossils have been found in the Middle East, Southern Europe, Eastern China and the Indonesian island of Java. *Homo erectus* had a bigger brain than *Homo habilis* and it was taller and possibly better able to move over longer distances. For the million or so years that *Homo erectus* appears to have existed, the basic

*Homo erectus* type seems not to have changed all that much, but the younger fossils of *Homo erectus* do have somewhat bigger brains than the older fossils.

The oldest fossils which indicate the next hominid type are around 135,000 years old and the youngest of this type are around 35,000 years old. These have been named *Homo sapiens* but subspecies *neanderthalensis* – better known as Neanderthal man. These fossils have been found in Western and Southern Europe and in the Middle East. Judging from the skeletons found of Neanderthals they seemed to have been heavier but not necessarily taller than modern humans. They had heavier brow ridges above their eyes and a big brain; in fact in many cases their brains were bigger than our modern average brain size. This is a great increase over the brain size of *Homo erectus* which is the most recent known hominid previous to Neanderthal man. The Neanderthal man fossils indicate that they were adapted to live in a cold climate and this makes sense as they lived during ice ages, Neanderthal man fossils only disappear from the middle of the last ice age onwards. There have been fossils found which are not as old as *Homo erectus* but are not like Neanderthal man either, and these are sometimes called archaic *Homo sapiens*. They look like an intermediate stage between *Homo erectus* and modern humans.

The oldest known fossils of *Homo sapiens sapiens*, or modern humans, were found in South Africa and are estimated to be around 100,000 years old. The oldest modern human fossils from Europe are only around 40,000 years old.

In this series of fossils there are many gaps, with names being designated to groups of fossils which have certain characteristics in common. As populations evolve, one species does not take over from the next in one generation, these changes happen much more gradually. What does happen is, that the rate of genetic change can be relatively slow for a period of time, and then go through a period of faster change. As genetic change and evolution are driven by changes in the habitat it would seem therefore that the environment also goes through cycles of more rapid change, interspersed with periods of relative stability. So what we hope is, that the groups of fossils to which we have given names are from individuals who lived during a period of stability in the environment. Up to the time of *Homo erectus* the environment we are talking about was that of Africa. From the time that *Homo erectus* moved out of Africa the changes in climate and habitat in Europe, Asia and the Middle East also influenced hominid evolution.

The next question is which fossil species developed from which, because both gaps as well as overlaps in time exist between species in the fossil record. This can make it difficult to conclude which species evolved from which predecessor. Yet, taking all the fossils and their ages into account certain trends are obvious. Our ancestors were similar to monkeys and apes and most likely lived in African forests. Then we get a series of fossils called the australopithecines which indicate that our ancestors started to walk on their hind legs only and this means that they had their hands free to carry things and use implements. Their brain was bigger compared to their ape-like predecessors, but the bigger change was in the body stance rather than in brain size. It seems that at this time in Africa a change occurred which resulted in the expansion of savannah and a decrease in forest area, and this seems to have forced the australopithecine ancestors from the trees onto the ground. The development of walking upright is an adaptation for living in open savannah land rather than for climbing trees.

Changes occurred amongst the australopithecines, and judging from the age of the australopithecine fossils it seems that *Australopithecus afarensis* gave rise to *Australopithecus africanus*, which in its turn may have given rise to *Australopithecus robustus*. Other australopithecine species may have also existed, but opinions differ about how they are related.

*Homo habilis* is recognized by many to be our first *Homo* ancestor. As already mentioned, some are of the opinion that this species should be part of the australopithecines. Opinions also differ about from which australopithecine species *Homo habilis* developed, some suggest that *Australopithecus africanus* was the direct ancestor, others think that *Australopithecus afarensis* was more likely. Suggestions have also been made that *Homo habilis* may have evolved from a species which existed previous even to *Australopithecus afarensis*. *Homo habilis* does seem to have lived during a time when *Australopithecus africanus* and *Australopithecus robustus* were also still alive, so perhaps it did evolve from *Australopithecus afarensis* or an even older species.

As with *Homo habilis* and the australopithecine species, *Homo erectus* and *Homo habilis* may have overlapped in time to a small degree, but most researchers seem to accept that *Homo habilis* gave rise to *Homo erectus*. *Homo habilis* did live during a period of great climatic change, with ice ages repeatedly advancing and retreating and maybe that is why both its origin and evolutionary destiny are still vague.

*Homo erectus* was the first hominid of which fossils have been found outside Africa. Its brain size and other characteristics do point to *Homo habilis* as its ancestor. The fossil record is, however, much scarcer for the time period between the youngest *Homo erectus* fossil and the oldest Neanderthal fossil. During this time the ice ages caused repeated periods of good climate alternating with periods of harsh cold climate in Europe and dry climate in Africa. This is also the time during which the two types of *Homo sapiens* evolved, with *Homo sapiens neanderthalensis* appearing in Europe and the Middle East and the first *Homo sapiens sapiens* appearing first in Africa and later in Europe. At one stage it was not clear whether Neanderthal man was the predecessor of modern *Homo sapiens*. These early modern humans are well known for the drawings they made on the walls of caves, one of the most famous examples of this cave art being discovered at Cro-Magnon in Dordogne in southern France. For this reason these early modern humans are often called Cro-Magnons.

Cro-Magnons and Neanderthals seem to have existed together in Europe for a period of time. It has been speculated whether Cro-Magnons either interbred with Neanderthals, violently eradicated them, or infected them with deadly diseases. What we do know is that Neanderthal fossils disappear some thousands of years after Cro-Magnon fossils appear in the European fossil records. On balance, the fossils seem to indicate that *Homo sapiens sapiens* developed in Africa and migrated from there into the rest of the world, like *Homo erectus* did before them. This means that Neanderthals were a side branch in human evolution. Whether Neanderthal man evolved from *Homo erectus* is not clear. Fossils of skulls deemed to be intermediate between *Homo erectus* and modern humans were found in Africa, Europe and Asia, and these are the archaic *Homo sapiens*. Modern humans may have developed from *Homo erectus* and if they did, they seem to have done so in Africa, rather than from *Homo erectus* elsewhere in the world.

Evolution and the development of species is fundamentally driven by changes in the environment, for this reason the most likely explanation of the emergence and subsequent disappearance of the various types of hominid fossils is most likely changes in climate. The ice ages waxed and waned, causing great changes in living conditions and food availability for all animals. Violence by one species towards another is part of most animals' lives. If you are a herbivore, you are likely to be a victim of aggression. If you are an omnivore or carnivore you have to be an aggressor in order to eat and you can also be a victim. Although it is possible for a predator to eradicate all of its prey, what is more likely to have such a devastating effect on an entire species is a change in habitat in combination with diseases and/or competition. The last ice age was not as severe as the previous ones and perhaps that helped modern man to spread further into Europe and put Neanderthal man, or its prey species, at a competitive disadvantage. For these reasons habitat change caused by climatic variations could be the root cause of the disappearance of an entire subspecies such as Neanderthals. However, we can only speculate about these prehistoric events in the light of our current ecological knowledge.

**The fact that early modern humans were named after Cro-Magnon is significant. The cave paintings which these early humans made represent one of our first known intentional efforts to store and transmit**

**knowledge other than by word of mouth.** We, hundreds of generations and thousands of years later, receive information from those Cro-Magnon people when we look at these images. Irrespective of what the particular immediate purpose was of these paintings, they still store and transmit knowledge information. We most probably don't understand the full intended meaning of these paintings, to do that we would need to know the Cro-Magnon 'code', just as we need to know the 'code' or meaning of any type of writing to understand what it means.

Our current ecology, reliant as it is on knowledge, is very dependent on the non-oral storage and transmission of knowledge because it is not possible to remember everything and communicate all our knowledge by word of mouth. **These Cro-Magnon people started a process of storing and communicating knowledge which has radically affected the character of human life, and underpins our current technological societies.**

Language and our voice, mouth, and larynx play a vital role in realizing the brain's potential in terms of communicating knowledge information. Our vocal cords and larynx are much lower in the throat of an adult human than they are in chimpanzees. Because of this, we are able to produce a much greater range of sounds than chimpanzees which are our closest living genetic relatives. Unlike chimpanzees, and because of the low position of our larynx, we cannot swallow and breath at the same time; if we do, then food and liquids go down the wrong way into our lungs and we choke and cough. Human babies can drink and breathe at the same time because at this early stage in our development the human larynx is still higher in the throat, like in other mammals. At about 1.5 years of age our larynx starts to descend in our throats, and by 14 years of age it has reached the position it has in adults. The australopithecines appeared to have had the larynx high in the throat similar to human babies, chimpanzees and other mammals. However, when *Homo erectus* fossils were examined there are indications that the larynx has started to descend. Some researchers have concluded that the position of the larynx in *Homo erectus* was similar to us when we are 6 years of age. The oldest fossils which indicate that the larynx was in the same position as our own, are between 300,000 and 400,000 years old and these fossil skulls are included in the archaic *Homo sapiens* group. These findings indicate that the australopithecines could not produce our human range of sounds. Human speech, probably together with our language ability, appears to have developed more or less hand in hand with the increase in brain size in our more recent ancestors.

We see the hominid brain starting its greatest increase in size, relative to the body as a whole, in the fossils which are included in the genus *Homo*. One assumption is that climatic changes in Africa, and the rest of the world, were the reason why we evolved our powerful brain. The repeated climatic changes brought about by the advancing and retreating ice ages happened, in evolutionary terms, in a relatively short period of time. Some researchers think that it may have been these relatively rapid environmental changes that accelerated hominid evolution. Obviously the changing ice ages did not happen at such a rapid rate that human genetics was not able to keep up. The combination of the timing of the ice ages and the ability of our ancestors to migrate as weather patterns shifted and habitats changed, may have been the reason why at this stage human evolution went so far and so fast compared to the evolution of other animals. We reached our present-day form in between 100,000 and 200,000 years, and this is much faster compared to the evolution of other animals. Our brain has up to the present been remarkably successful and has given us very powerful advantages, and these advantages have forced our evolution forward so fast.

One of the far-reaching effects of the increase in human brain size is the way it influenced our reproduction. The brain size at birth of most of the larger primates, except humans, is about half that of the adult brain size. A human baby's brain, on the other hand, is only between a quarter and a third the size of an adult human brain. For a human baby to have half the size of the adult brain, we would have to have a 21-month pregnancy. Because our brain is so big relative to our bodies, our babies have to be born while their heads still fit through the human pelvis. A baby's head would be too big if we were born at a later stage in our development. Our human pelvis is relatively constricted compared to that of other primates because it needs to be a certain shape to allow us to walk upright. We are the only primate that walks fully upright.

A baby chimpanzee is actually more developed compared to a human baby when it is born. It has half the brain size of the adult chimpanzee and is more capable at birth than a human baby. The human baby at this stage only has less than a third of the adult's brain size. That is why human babies are so helpless and need full-time care. However, after a number of months a human baby quickly surpasses a baby chimpanzee in what it is able to do.

So the increase in our ancestors' brain size, in combination with changes in our pelvis due to walking upright, has forced babies to be born at an increasingly earlier stage in our development. This also means that our babies and children need more intensive care, and for much longer, than the offspring of our closest living genetic relatives, the apes. The time it takes our brain to grow, and the time which is needed to teach us all we need to know are the longest of any primate. All this is as a result of the large size of our brains. **The rearing, care and education of our offspring are essential parts of our ecology and our survival, and takes approximately 20 years to**

**complete. This whole process, without which we would go extinct, comprises human reproduction, not just sex and a 9-month pregnancy.**

In the past we did not have the information about our human origins, and many and varied ideas and myths about our origins existed and were also part of religions. We can now consult the physical evidence such as fossils many times over, repeat experiments and debate the various interpretations of our origins. This is why conclusions about our origins and ancestry included in The Natural Religion are based on fossils and biochemical information, as this is the most reliable evidence available to us at this time. Our knowledge about our origins based on fossils has many questions and gaps, and it is not entirely clear what our exact line of descent is; but the fossil record does give us an insight into the process by which our species came about.

Fossils are not the only source of knowledge about our origins. Information about the way our bodies function, our biochemistry, physiology and genetics, also provides us with information about where we came from. The species of apes which are alive today only differ a few percent from us genetically. The ancestors which we have in common with the present-day apes existed before the australopithecines, so we are presumably more closely related to the *Australopithecus* and *Homo* fossil species than we are to the present-day apes. Examination of mitochondrial DNA points to Africa as our place of origin, and other human biochemical characteristics of people from all over the world also indicate that modern humans all originated from one region and then spread out over the rest of the world comparatively recently. All this evidence supports the theory that our fossil record represents parts of what is a progression of related species. **Our brain has made us a very special animal species, it has given us the power to inhabit most of our planet, and is most likely the reason for the unusually fast pace of our human evolution compared to other animals.**

Traditionally religions have explained where humans come from in terms of one identifiable beginning rather than a gradual process. However, **the knowledge-based approach of The Natural Religion sees a gradual evolutionary development as the origin of the human species.** We did not suddenly become 'human' from one generation to the next, in evolution species do not suddenly change overnight, what happens is a much more gradual process. What we call a species is just a series of generations which do not vary a lot from each other. Human beings gradually emerged, gene by gene we evolved more and more into what we now call being human. This means that if we now call ourselves fully human, that our evolutionary ancestors were not completely but partially human. It also means that if our species continues to exist, future generations, in some hundreds of thousands of years' time, may not view us as completely human. They may think that we are only 75% or 80% human, just how we think our ancestors were. Alternatively, they may still calls us human, but call themselves something else.

## 4.3: Our ecology

Our ecology is dominated by the tools that we use, and by the technology that we produce, which we have because of our brain. Many non-human animals also use tools. Finches, vultures, otters, dolphins, monkeys, ants, all use certain objects in their environment as tools. Chimpanzees even modify objects to suit the particular task at hand. Tools made of flint and bone have been found together with the fossil remains of our australopithecine ancestors of 2.5 million years ago. As the hominid fossils become more recent, the tools that are found with them become more sophisticated. *Homo erectus* also had tools, but these were better finished than the australopithecine ones. They appear to have had spears and other implements with sharp points and edges, and used snares to catch animals. The remains of *Homo erectus* in China were also found together with the remains of fire. So, these pre-humans apparently knew how to control and use fire. These tools and techniques allowed our early ancestors to access resources that would not otherwise have been available to them.

So, the more the brain developed, the more tools were used, and the more complex and powerful the tools became. Of all animals, we have the most powerful tools, and rely on them most in our everyday lives. In fact, tools are essential to our current lives and ecology, of all animals our ecology is the most dependent on tools. Our use of technology and tools mirrors the development of our brain and is part of a continuum which started a long time ago in the history of animal life. The animal brain is a living organ, which is part of a living being, which is part of the biological world. What it produces is a product of life, therefore tools and technology are a natural product of life.

As we live our everyday human lives we tend not to think about how closely we rely on our technology for every aspect of our lives. Our clothes are a type of technology or tool to keep us warm and protect ourselves, and our shelters or dwellings are technology which provide us with safety and shelter. Without shelter and clothes no modern human could live in temperate or colder climates, because we are essentially tropical animals. Our food is produced by means of technologies, tools, and techniques. Even those very few of us who still live entirely by means of hunting and gathering, use tools to gather and store food. **Our technology and tools are not an 'unnatural' aspect of our lives, nor are they an 'artificial' part of our ecology. Tools and**

**technology have been part of our ecology and evolution even before we call ourselves human, they are part of the reason why our evolution moved ahead so fast.** The Natural Religion sees tools and technology as part of our natural lives just like the legs of a horse, the wings of a bird, the claws and teeth of a lion, or the fins of a fish. Technology and tools transform knowledge information in our brain into tangible resources.

Apart from one major attribute, we humans are pretty mediocre animals. Except for our brain, other animal can do most things better than we can. We humans are what are called 'generalists'. This means that we can do a wide variety of things, but none of these we can do as well as the specialists in that particular activity. We can walk and run, but even a small dog can outrun most people. We can swim, but many mammals can swim faster or further than humans, not to speak of fish. We can climb a bit, but our climbing does not compare to animals such as mountain goats, squirrels and monkeys. We can see, hear and smell, but other animals have much greater acuity in one or more of these senses. We are to a certain degree resistant to diseases, but animals such as sharks and rays, who already looked like they do now at the time of the dinosaurs 65 million years ago, appear to have developed superior disease resistance than we have. However, animals who are more specialized in one ability tend to have limitations in others. For instance a horse or fish cannot really climb, and mountain goats and monkeys are not good swimmers.

In combination with our very special brain, we humans have quite a wide range of abilities, but although wide ranging most of these are mediocre. Being generalists has been a great advantage in combination with technology and tools produced by our brain. Using knowledge we made tools and technology which allow us to run, swim and fly further, faster and higher. We have even created ways of increasing our disease resistance and preserving our health with our medicines and medical know-how, all this is part of our human ecology. We depend on our brain to give us abilities that the rest of our body does not have. This is not unnatural or artificial, but has been part of our species and the species from which we evolved, for tens of thousands of generations. Tools have helped to spur on our evolution.

Previous to the time for which we have historical records, we can only infer from other evidence what our lives and ecology were like. From the flint tools and bones found with our *Australopithecus* ancestors we conclude that hunting was part of their lives, but their teeth indicate that vegetation, fruits, roots and the softer green shoots made up most of their diet. We presume that they lived in social groups of individuals who were related to each other. We have evidence from sites in Africa and China that *Homo erectus* appeared to live in groups. What the world population of Australopithecines was is not known, but one estimate of the world population of *Homo erectus* at about 1 million years ago is 125,000 individuals. This has been estimated to have risen to 1 million by 300,000 years ago. It is during this time that *Homo erectus* spread from Africa to Asia and Europe, and this most likely led to the increase in the world's *Homo erectus* population. Human world population estimates rise to 3.34 millions by 27,000 years ago, and to 4 and 5 millions by 12,000 and 10,000 years ago respectively. However, estimates rise to between 27 and 70 million people worldwide by 4000 years ago, and between 133 and 255 millions 2000 years ago. These estimates indicate a sudden sharp rise in our world population in a relatively short time from approximately 8000 years ago onwards. Such a marked increase in population numbers of any species is usually a sign of an important change in the ecology of that species. So what happened in our human ecology around 8000 years ago?

The ecology of the *Australopithecus* and *Homo* species consisted of what we now call hunting and gathering, groups of closely related individuals moved together from one location to another in a relatively large area. The movements of the group would depend on when certain trees had ripe fruit, or when tubers and roots of certain plants were fully developed, and when seeds and grains were ready to be gathered, and also on the migrations of the animals they hunted. Water supply would also have dictated when a group would move from one area to another. Typically, there may have been between 15 and 40 people in a group. The size of the area they depended on would have varied according to how rich its food resources were, but to give an idea, hunters and gatherers of recent times only live in densities of between 0.1 and 1 person per square mile (0.06 and 0.6 person per square kilometre). These are very low population densities compared to the way most people live today.

The groups in which our ancestors lived would have consisted of parents and their offspring of various ages. Either the young adult females or males would have moved from the group they were born in to another group to avoid inbreeding. In this type of ecology the possessions of a person could not be much more than what they could carry with them as they moved through their territory. So, this not only meant that possessions were kept to a minimum, but also, that a mother could at any one time not have more than one baby or young child, because it had to be carried. As their low population densities suggest, our ancestors probably did control their fertility. One way that helped to keep the number of children low that women had during their lives, was most likely by breast-feeding babies and toddlers for years rather than months. It is probable that family size was also limited by other means.

Our knowledge of current hunter-gatherer groups indicated that food resources, particularly food from plants, tends not to be in short supply. The members of a group tend to be healthy and can easily gather enough plant food to eat. In fact, some present day hunter-gatherers generally don't store food despite knowing how to do this, because they can get enough fresh food at any time. The reason why current hunter-gatherer ecology is like this, is that their population densities are low and continue to be low. Our ancestors of hundreds of thousands of years ago are thought to have had a similar ecology to that of our present day hunter-gatherers, so it was most likely this type of ecology and life style that genetically formed us.

During the relatively short period in terms of the history of our species, between 10,000 and 4000 years ago, our world population suddenly grew much faster. **This jump in population was the result of a major change in our ecology, which was that we began to grow plants and rear animals instead of gathering and hunting them in the wild. Humans started to practice agriculture.**

Agriculture has changed human life in very many ways – changes which most of us now take for granted.

Growing and rearing plants and animals was carried out by other animals long before we did. Many species of ants tend and herd aphids (greenfly) and other insects, and gather honeydew from them. Some species of ant even thin out some of their herd of insects and eat them, as well as eating the honeydew they produce. This is very similar to our own use of farm animals. Culturing food to support large populations was evolved by approximately 40 species of two genera of leafcutter ants in North and South America millions of years before we invented agriculture. They grow a fungus on clumps of chewed-up leaves and then eat this fungus. The practice of this type of agriculture is based on genetic information and it is very efficient as the success of social insects shows.

Our own agriculture is based on knowledge information. We have learned the techniques to grow plants and to rear animals. If, for some reason, we were to lose all our knowledge about agriculture, we would have to experiment and develop it from the very start all over again. The ants species that practise agriculture do so because they are genetically programmed to do this. We humans are not genetically programmed in this way, but although we developed agriculture using knowledge information while the ants used genetic information, the advantage to both our ecologies is great because it is a very efficient and reliable way to produce food. Gathering plant food from the wild and hunting animals is much less efficient than agriculture because the way wild plants and animals grow is not directed by us for our benefit. Wild food sources are impacted by weather events such as drought and frost, other animals which eat them, diseases and parasites, and competition from weed species. If we culture plants or animals, we can protect them from diseases, eliminate other plants and animals from competing with them, and stop other animals eating them. In other words, culturing plants and animals maximizes their growth potential even before we also alter them genetically by selective breeding to suit our purposes even more.

Human ecology is greatly influenced by the increased and more predictable food supply which agriculture gives us, but it also influences our ecology in another way. When culturing plants and animals this is usually done in a certain location for at least one season. In fact, agriculture has in some cases been carried on in the same location for over one hundred human generations. This has meant that people who practise agriculture do not travel about anymore, but stay living in one location. If you live in the same place, you can accumulate heavy tools and implements. Hunter-gatherers cannot do this because they can't take them along when they travel to a different location. So the life-style which goes with agriculture allows people to accumulate technology.

A result of the increased production of food is that a person can harvest more food than he or she needs for themselves. This means that some people don't have to grow food for themselves, as they can eat the excess of others, therefore these people have the time to do other things. This has had a great influence on our ecology because people that have the time to do so, can create and invent new technology, they can experiment, examine and reflect, thus conceiving and gathering knowledge. This knowledge can then be used to improve agriculture further, thus increasing food production, and releasing more people to do other work. One of the conclusions of The Natural Religion is, therefore, that our agriculture is the basis of modern human life and its knowledge boom.

A negative side of the accumulation of food and other possessions and resources is that they attract other people to come and steal them. If one group of people happen to have a shortage of food and others have a surplus stored away, this creates a great attraction for the group that is short of food. This is a basic situation which occurs in the ecology of all plants and animals. An unequal distribution of resources, or resource gradient, results in the build-up of stresses between members of the same species and at times also between different species. These stresses are every bit as much part of our present-day ecology as they were at the time when agriculture first developed. As far as we know, it was when agriculture developed that large-scale warfare between groups of humans arose. The Natural Religion's attitude towards war is that it is an extremely wasteful and unsophisticated way in which to solve the problems caused by unequal distribution of resources. Unfortunately war has been an integral part of human history, certainly since the advent of agriculture.

Both the way we distribute our resources and warfare are discussed in more detail in Chapters 8 and 9.

People who are released from having to grow their own food and therefore can do other work, can specialize in doing one specific type of work. This means that they can become very practised and efficient at doing that work. As a result there emerged societies permanently living in one place, with people specializing in various tasks. Because there was enough food, and because mothers did not have to carry babies and young children over long distances, family sizes grew, and therefore so did the overall populations. This is the way villages, towns, and cities first developed.

In urban areas people tend to do one particular type of work and depend on others to provide them with their other needs. This has created highly integrated societies, in which people depend on other people to provide them with most of their daily needs. This means that when one lives in an urban society you are very dependent on other people. This type of society has become more and more important in human ecology. The reason for its success lies in the fact, that when people specialize in one or two types of work, they become more efficient, compared to a person who does many different tasks. Urban society is really a large organized group of people all carrying out one or two specialized functions.

Again, we humans are not the first to have developed large organized groups or cities with individuals carrying out specialized tasks. The social insects developed specialization and what are effectively urban groups before we did. As with the development of agriculture by ants, the reason why ants, bees, wasps and termites developed their organized colonies was the same as the reason why we developed our urban societies. Social insects, as well as us humans, can support more individuals in this organized and specialized system because it is more efficient. **As with agriculture, the social insects developed their colonies and nests based on genetic adaptation, while we developed our urban societies through the use of knowledge.**

The single most important effect which the development of agriculture, and the efficiency it brought with it, has had on our ecology is the phenomenal rise in the world's human population. Particularly when we compare it to the estimated rises in population of our ancestor species, and the short time scale in which it took place. Our increased population densities have had fundamental and far-reaching consequences on our ecology and for this reason Chapter 7 is devoted to examining the practical consequences of our growing population. For any living species high population densities fundamentally affect their ecology. For example, as mentioned above, the development of specializations is one way many species deal with this. Special ways in which resources are divided out, and social relationships, also become of prime importance. Food shortages, diseases and aggression towards members of the same species, are very important in the ecology of species with high population densities. All of these factors became pressing matters with us also, when we began practising agriculture and our population densities increased. What may have been minor issues in low-density hunter-gatherer societies, became major problems in high-density agricultural and urban societies.

The human knowledge revolution began at the time we started practising agriculture 10,000 to 12,000 years ago. Then, our species had already existed for approximately 100,000 years. Our rise in population has given us an important human resource that fuelled our huge advances in knowledge. More people create more knowledge and technology, because knowledge and technology do not come about by themselves, people produce them. Our increased knowledge in its turn has had the effect of increasing our populations even more. It was knowledge which further developed agriculture, as well as creating and improving other technologies which contribute towards the health and welfare of people.

Agriculture is therefore an absolutely essential part of our current ecology. We could not support the present-day world human population without it. It is because of the way that agriculture produces food that our world population has been able to reach this level in the first place. This gives some idea of the extent to which agriculture changed the ecology of our ancestors and is the base of our current ecology.

In historical times, dominant political powers were based on agriculture. So also the fall of political empires went hand in hand with failure of agriculture due to changing climate, erosion, or exhausted soils. Our present-day societies continue to depend on the culturing of food. Even those few groups of people who still have a predominantly hunting and gathering life style usually trade with, and sometimes work for, people who grow crops, as in the case of Pygmies in the Central African rainforests. Presently in more fertile and developed regions, agriculture is capable of producing many times the food requirement of the people who actually grow the crops and tend the herds. This means that the majority of the population have time to do something else, which has allowed the industrial revolution to happen. Again, specialization in industry gave rise to increased efficiency with less people making more products.

Agriculture has also led to our use of money and the economy which is based on it. People who specialize in growing food, or making tools and implements, or providing a service; need to exchange what they produce for the other things they need. Farmers exchange food for tools and implements, and service providers help farmers and tool makers in exchange for food and implements, so also with the tool makers. This system of exchange gave rise to money which does not have its own inherent value, but is essentially a universal exchange medium. The real value of money is what it represents rather than what it is. This whole system of production and exchange has become an important part of our ecology. Food, tools, implements, services and money are all part of our human system of resource partitioning. This exchange system is called our economy.

As we will see in Chapter 8, the welfare and quality of life of the majority of people on this Earth depends on our economy. We all need resources to live, and most people do not produce all the resources they need themselves. **So our present-day human ecology is characterized by exchanging goods and services. This means that virtually everybody on Earth is involved in the human economy.**

Part of The Natural Religion is therefore that the fundamental purpose of life is to capture and store energy, and the mechanism by which most humans do this in our present ecology is via the medium of our economy.

The industrial revolution took place in some regions of the world because agriculture gave us the time to do this. Non-agricultural products have increasingly become a greater part of our economy. Many of these products are essential to our lives, but others are non-essential and are classed as luxury goods. Together with the industrial revolution, the service industry has also grown primarily in the currently richer parts of the globe. Since historical times, there have been resource gradients between human populations living in different parts of the world. Regions grew wealthy and then diminished in wealth again, and this can be described as the resource gradients steepening and subsequently evening out again and occasionally reversing. Since the advent of agriculture, and up to the present day, steep resource gradients have existed in our ecology, meaning that there are very rich and very poor people in our communities. This sets up social stresses.

The effects of poverty in parts of, or whole, societies have far-reaching detrimental consequences on people. Worldwide, the less wealthy you are, the less healthy you tend to be. Poor societies or sections of societies also underachieve because they lack investment in the resources that can produce wealth, resources are needed in order that people can realize their potential more fully. This means that society as a whole does not benefit from the full potential of human effort if many people are poor. Wealthier sections of society are able to help poorer sections. If this help is not forthcoming, the inequality of resources will be seen as unjust and unfair. Thus resentment builds up, causing jealousy and stress between people, which has the potential to result in physical violence, as we will see in Chapter 9.

The prominence of violence and war in human ecology since the advent of agriculture should not be surprising. Aggressive competition is regularly associated with high population densities in the ecology of many species. So also is aggression associated with steep gradients in essential resources. For instance ants, which, as we have seen already, also have agriculture and live in organized colonies or cities, go to war with other colonies of ants in order to steal from them. However, since we have evolved the capability of using knowledge, we now have the potential to solve these problems in more sophisticated and nonviolent ways. Other animals have used genetic mechanisms in their ecology to also solve these problems in non violent ways. Inhibitions in behaviour exist when certain animals fight amongst themselves, so that they don't actually seriously harm or kill each other. For example, wolves can act like this. There are many examples of animals intimidating each other just by sound, visually, or with smells; as in birds, fish and cats respectively. Many species avoid physical fights at all costs, unless placed in abnormal and extreme situations. Other species have evolved not to fight but to display to each other. Still others fight in ritualized ways, which may test each others strength, but the contest is essentially non-destructive.

**Some human societies can maintain their social organization and structures without the overt threat of violence, certain hunter-gatherers and small village-based societies are examples. Therefore, the opinion included in The Natural Religion is that while we have many wars, this is not inevitable, because we know that we can have virtually violence-free societies.** As mentioned above, Chapter 9 looks at this in more detail.

We have seen how agriculture gave us time to gather knowledge, which has metamorphosed into our powerful technology. Of course, our present day ecology does not depend on technology in all regions of the world equally. Cities and urban areas depend almost entirely on technology, but we also live in small scale subsistence farming societies where agricultural techniques may not be all that different from those of thousands of years ago. Yet, virtually every region of planet Earth is in one way or another influenced by our technology. Even in mid-ocean, satellites orbit overhead, technology-generated pollutants are in the water and the air, and planes and ships pass by. The power of our technology has spread all over the world.

Nevertheless, we are part of the living system on Earth. The Natural Religion points out that we depend on other living species such as our plant crops and farm animals to help us stay alive because technology has not yet given us the capability to exist independently of other life on Earth. **Knowledge is the hallmark of human ecology, and despite the obvious changes which the industrial and technology revolutions have brought to the way we live, the agricultural revolution still remains the one that so far has had the most fundamental effect.**

How will knowledge information effect our future ecology? We hope that knowledge itself will tell us that, and help us solve whatever problems our knowledge-ecology may have. One characteristic of knowledge information is that it can change and act very quickly compared with genetic information (see Appendix), which means that the changes to our ecology caused by knowledge also occur fast. Our knowledge is growing, therefore its effect on our ecology will increase in the future.

The word 'ecology' is essentially a scientific term and could be seen as remote from our own experience of life. It is after all through our thoughts, feelings and emotions that we experience our lives. However, the description of our personal lives in this book is drawn from our current knowledge of human life, which, in fact, is our ecology. We cannot escape our biological needs, and the fact that we are part of Earth's biological system. For instance, the kind of life any one person leads is very much influenced by the resources they have. Food, fuel and shelter determine whether a person is rich or poor, has a good or bad quality of life, is healthy or sick, or has the luxury to choose from a range of life-options. By focussing on the real basis of our existence – in other words, on our ecology – The Natural Religion hopes to help people gain a better understanding of how human life works.

For instance, and as we already saw, our knowledge and technology are not in any way 'artificial' or 'unnatural', they are the natural products of our human brain and therefore biological phenomena; but we should realize the extent to which knowledge has transformed the way we live, and how fast this has happened. An important implication of all this is that in order to guide our lives and our ecology in the right direction, we must actively use our knowledge of ourselves to achieve this. **Therefore, part of The Natural Religion's outlook is that we humans as a species do not have an automatic safety net, we need to safeguard our future lives and our future ecology ourselves, because this will not be done for us.**

## 4.4: The Human Evolutionary Experiment

Advances in technology within the last 150 years happened more rapidly than any of the technological advances over the thousands of years before that. As mentioned above, we therefore have to rely on knowledge information to deal with the fast changes that our ecology is now going through. To rely on adapting genetically will almost certainly be much too slow to keep up with the fast pace of changes in our present day ecology. Knowledge is having a fundamental influence on the way we live and is doing so faster and faster.

**We humans are living an evolutionary experiment. An experiment whereby genetic information has created knowledge information.** An experiment about the effect of knowledge information on the ecology of a mammalian primate species, namely us human beings. So far human knowledge has been extraordinarily powerful. Through our knowledge we have achieved success in less than 200,000 years, which is an extremely short time span compared to the time it took other species to evolve.

Whether this evolutionary experiment will continue to be a success, we don't know yet. As far as we know we are the first species that ever evolved knowledge to this level of understanding; having as much knowledge as we have, has never been tested before. It is possible that despite our success to date, the power of knowledge could be detrimental to us as a species in the long run. Our boom success story could be followed by a bust. A population explosion followed by a collapse is something which is a regular occurrence in the ecology of living species. Large swings in population numbers occur in many animals, ranging from locusts to lemmings, rabbits, and deer. Yeast cells in beer or in wine first grow vigorously whereupon their population collapses in the brewing vat due to having used up the sugar they feed on and being poisoned by the alcohol they produced as a waste product. This is a typical and basic sequence of events in any population of living organisms.

Human history is littered with examples from all continents of famines due to human populations outstripping food supply. Our technological know-how has produced nuclear weapons that have the potential to kill millions of people and make the world uninhabitable for thousands of years in the future. So we don't know if our experiment in the development of knowledge, with ourselves as the guinea pigs, will continue to be a success. It will depend on the way we use our knowledge information to direct our own future ecology.

**One of the aims of this book and of The Natural Religion is to help us use our knowledge to make informed decisions, safeguard our ecology and our species, and improve our quality of life.**

## 4.5: Essentially Human

Opinions about what makes us essentially human, and the way people see themselves with respect to all the other living beings around us, varies greatly at different times and in different cultures. Some cultures see ourselves as being separate or apart from nature; as fundamentally different from all other life. One reason for this is the belief that we have a supernatural part to us and all other life does not. Other cultures take the attitude that we have a spiritual side to us, but that the spiritual world also includes other living beings such as animals and trees.

As we know, the ideas in this book are based on our knowledge. No verifiable knowledge exists that we possess a supernatural or spiritual side. Knowledge, as mentioned before, deals with the natural world and therefore cannot check, examine, question, or verify anything that is not part of the natural world; including which, if any, of the different supernatural ideas and concepts suggested all through the ages by people in different cultures all over the world, are correct.

We see ourselves as being human; so depending on how long ago our evolutionary ancestors lived, they were partially human to varying degrees. We developed gradually by means of the process of genetic evolution by natural selection; there was no one point at which we were suddenly human, and before which we were not.

The overall rate at which humans evolved appears to have been very fast compared to other species and was most likely due to the powerful advantages which our brain gave us. Of all animal species, we have the most developed brain and are capable of the highest level of understanding. Our brain and its capabilities is the single most important factor which makes our ecology so unique compared to that of other animals.

Having the most developed brain of all life on Earth does not mean that we are not still in every sense part of the Earthly living system. Many plants and animals are the most developed with respect to certain particular characteristics. Certain trees are the heaviest or oldest living species on Earth; but just because species may be faster, taller, heavier, or older than the others, does not mean that they are not part of life on Earth. We humans happen to have the most developed brain of all animals, but this does not mean we are inherently different from all other life. However, it does mean that we are self-aware and that we have the mental capacity to compare ourselves to other living things. The fact that we can do this is a human quality, but does not mean we are not in every respect part of the Earthly living system. The extent of the understanding of which we are capable – the thoughts and emotions, and appreciation of the significance and implications of events – are all part of being human.

In many respects our ecology is like that of no other animal, it is uniquely human. No other living animal species has changed its ecology to the extent that we have, our impact on our world has been extensive and our essentially human influence can be seen all over this planet. Some of these influences are good and some are not, but we are still an integral part of the living system, and need it to survive. Part of The Natural Religion is the warning that if we damage parts of life on Earth, it may result in serious consequences for humankind. However, our brain can understand our own power and its good and bad effects. Therefore, we also have the potential to use our knowledge for the good of our ourselves and the environment. **The conclusion reached in The Natural Religion is that our brain is the essence of being human.**

One luxury that our human brain gives us, is the ability to make choices, both correct ones and mistaken ones, and the implications of this we now discuss next.

# Chapter 5

# Our Personal Existence and Our Choices

🌿 We influence our own personal existence 🌿

## 5.1: Our personal being

Of all the things that we are aware, we are most aware of ourselves. The sense of our own being is probably the strongest emotion we have. Our sense of 'personal being' feels private to each of us. We all have feelings and emotions that we would prefer not to reveal to anyone. We sometimes think that our existence is unique, we can feel distinct and separate from all other people.

Religions help us think about who we are as individual people, as well as our position in relation to those around us. Some traditional religions state that the essence of a person is their supernatural soul or spirit. As we have already concluded in previous chapters, we cannot verify one way or the other whether this is actually the case. Beliefs about a soul or spirit differ in various parts of the world and many contradict each other. A person's particular belief or faith depends largely on where and when they were born.

The approach of The Natural Religion is to look at what we are physically. We are the combination of the thousands of trillions (a trillion is a 1 followed by 12 zeros) of living cells that make up the skin, bones, muscles, organs and blood of our bodies. This number is so large that we cannot comprehend it. As biological beings we are so complex that it is difficult to fully understand it, but we tend not to think of ourselves in terms of the detail of our full biological complexity. Instead, we see ourselves as whole living beings; we test what we can and can't do, we study how we differ from other people and we monitor how we feel. However, in recent times we have gained so much knowledge about ourselves that we understand what life is and how it works in our bodies. We now also know that we have feelings, thoughts and emotions because of the way our brain functions and that none of these human characteristics need any supernatural explanations. **Therefore the basic standpoint included in The Natural Religion is, that we are biological entities, existing because of highly complex biological structures and reactions of life, and that we experience our own being with the thoughts and feelings of our brains.**

Our thoughts and feelings about ourselves normally compare us to other people and our general surroundings. No living organism exists in a vacuum. All of life lives in and interacts with its physical and social surroundings. We are a part of the human species and our individual actions all contribute towards what humankind as a species does. Together with the rest of biological life, we are part of planet Earth. However, most people's immediate concerns are, and need to be, about themselves because if we don't survive as individuals, then humankind as a species will not survive. So we have evolved to look after ourselves, as well as to be part of a species as a whole. This also means that, while we depend on and help each other, we also compete with each other. We experience the world from the perspective of our own separate personal existence. So, also included in The Natural Religion is that while we experience our existence with our brain, **what we experience as our personal being is a combination of our own bodies interacting with the physical and social circumstances in which we find ourselves.**

## 5.2: Making choices

The evolution of nerves and a brain in animals gives the ability to make choices. The human brain can gather, hold and process knowledge information so we can evaluate options and choose the one we want. In contrast, and as explained in the Appendix, genetic information does not allow choices to be made during a lifetime. Our genes form a blueprint which is fixed from fertilization onwards. However, as mentioned in Chapter 4, genetic evolution has managed to get around this inflexibility by developing nerve and brain tissue that allows choices to be made.

We have to make decisions most days of our lives, and we make them in many different ways. Some decisions are well thought out and aimed at a specific goal, others are spur-of-the-moment decisions and still others intuitively just feel like the right thing to do. Choices are sometimes made to attain a short-term goal such as immediate happiness or gratification without necessarily taking the longer term consequences into account. We humans try to guide our own futures and create a significant part of our own lives with our personal choices.

We cannot see into the future, nor do we always have all the information that we would like to have when making decisions. For these reasons our decisions don't always result in the outcome that we hoped for and therefore we can make mistakes. Few, if any, people exist or ever existed who never made a mistake in any of the decisions they took during their lifetime.

Since we cannot see into the future, the only way we can try to ensure that the choices we make are good ones is by using the knowledge information available to us. Good knowledge is crucial, because we must have knowledge to make good choices and to plan and predict the future.

Knowledge information gives us choices. Animals whose ecology is based more on genetic information do not have the level of choice that we have. Social insects, such as ants for example, cannot make the choices that we can. The way they find food, make their colonies, reproduce and their whole ecology is under more direct genetic control. Because of our human brain we have the greatest freedom of choice of any animal species on Earth, and the decisions that we make fundamentally influence our ecology. **The Natural Religion aims to help people with the choices we make in life by stressing the importance of knowledge in making good decisions.** Our choices help to shape our own lives and we need knowledge to do this.

## 5.3: Making errors and mistakes

As mentioned above, one drawback of having choices is that we can make bad choices and decisions. When people make a bad choice, it is a mistake or an error. The term 'error' and 'mistake' in the context of this book are used in the very broadest sense of these words, in that they can have unexpected or bad effects on the person that made them and/or on other people. Traditional religions have given specific names to choices that are deemed to be wrong and that they don't want people to make. In the case of the Christian religion, the word 'sin' is used for this and carries with it the meaning that it breaks the rules of a supernatural deity. In The Natural Religion the words 'mistake' and 'error' are used for wrong actions and for bad choices in order to stress that they have bad practical consequences. For example, stealing takes away resources unfairly that the owner may be depending on, and murder cruelly and tragically takes away our very existence. These are very serious errors and mistakes, with such bad practical consequences as suffering, grief and pain.

Choices in many aspects of our lives may have lesser or greater disadvantages associated with them. As we are part of the ecological system, most of our choices involve striking a balance between two extremes. In real situations, we have to try to assess the best balance for the circumstances we are in and on the basis of the knowledge that we have. To be realistic, it is unusual if we can make a choice that is ideal in absolutely all respects. Nevertheless, we can make decisions that have many good and few bad results, and these are the decisions we normally attempt to make. We also have choices that directly or indirectly, have mostly bad consequences and these can range from small mistakes to very serious errors.

People can make mistakes inadvertently by making a decision that results in a completely different outcome than was expected. A person may plant a crop and the weather may suddenly change, so that all the young seedlings are killed. A person may decide to live in a certain place, but this place may subsequently prove vulnerable to flooding, earthquakes, diseases or attacks from other people. A person may join and cooperate with others, but these people may not act or reciprocate in the way that the person expected. These types of mistakes happen because a person may not have enough information or the information may not exist to be able to make a good choice.

People can also make mistakes by taking the wrong decisions even though they do have the knowledge to make better choices. One reason why people make such avoidable mistakes is that they do not think deeply enough about the consequences of what they are doing. For example it is often easier to consider the benefits of a decision just for oneself and not to think about the possible harm it could do to other people. Such choices are mistakes. We are a social species, so we have to live together and in cooperation with other people. Decisions that are too close to the egoistic extreme, and ignore or actively harm the interests of other people, are mistakes or errors.

We humans have created social systems that reconcile the sometimes conflicting interests of the individual with those of the community as a whole. We all benefit from being part of a community, and what a community is like is due to the way we act towards each other. An important function of a social system is to determine where the balance lies between protecting the rights of the individual and the interests of the community. It draws up a set of rules or laws that indicate what is deemed to be fair and unfair. These laws help to create dependability and predictability in a society. We all need to have food, drink, and shelter; we need to meet and keep sexual partners, maintain our relationships and social bonds, and stay healthy and safe. Dependability and predictability in our social lives help us to do these things, and people have created this for each other. We all contribute towards making the lives of others in our community dependable and predictable. Not caring for oneself is a mistake, but so also is not contributing to the community or giving ourselves an unfair advantage over others.

For example, to harm or kill another person is wrong and a very serious error. People who have killed other people, as well as causing suffering and grief, accept and rely on the protection of society at large to stop others from murdering them. Also, to steal from another person is a mistake and very unfair. Thieves hope to steal from others but rely on the rest of the community not to steal from them.

When we make these kinds of mistakes and errors we usually hope that we will gain from them and that the bad consequences will only affect other people. However, all of us together make society what it is. All of us influence our social environment to a lesser or greater extent. So even if we are not directly affected by the bad consequences of a mistaken decision, we still have to live in a community that is adversely affected by our own errors and, therefore, indirectly harm ourselves.

A person who acts unfairly towards others may either not realize the full consequences of their mistaken choices or may simply not care about or ignore them. Although it is difficult to act unfairly towards others and avoid also adversely affecting oneself, in certain situations this may be possible in the short term. However, it is very much in the interests of a society or community as a whole to try and stop unfair behaviour. Both on a personal level and as a community, we tend to be very sensitive about being treated unfairly by others. It is essential for any group of people that their society attempts to stop people acting unfairly.

What is thought to be 'fair' and 'unfair' may not be the same in all cultures and societies. The balance struck between an individual's rights and the interests of the community can vary. A compromise has got to be arrived at between what is the extreme ideal for an individual and the extreme ideal for society. As with most extremes in ecological systems, both are undesirable. The task of determining the compromise that allows a society to operate successfully is usually up to the social systems of that society, its civil laws and rules, and often also the moral codes of its religions. The Natural Religion recognizes that we are all capable of making mistakes in decisions that are not well thought out or simply too selfish, and that the basic rules and laws of most societies attempt to stop us making those errors.

Our most important possession is our life. To preserve and maintain it is our most immediate and personal purpose. There is no more fundamental way in which to affect a person than to take away their life. **It is wrong and one of the gravest errors that we can make to take away someone else's life either intentionally or unintentionally. It is a wrong that we cannot**

**rectify because we cannot, as yet, give life back to someone who is
dead.** Life is very precious, which is one of The Natural Religion's most
fundamental ideas.

Unfortunately situations exist in which the rules of many societies do allow
one person to kill another. If the extreme situation of one person attempting to
kill another occurs, and the threatened person is absolutely sure that their life
is in danger and they have no other option to save their life, then he or she is
usually allowed to defend themselves effectively, which in certain situations
could potentially result in the death of the aggressor. The principle of self-
defence is part of the living system: every living species has developed ways in
which to defend itself. The problem with self-defence leading to someone's
death is how to assess that the life of the person who is being threatened is
really in danger and that they have no other options? If a person fears for their
life they have to make up their minds in that stressful, and probably completely
unfamiliar, situation whether they have no other option but to defend themselves. In addition, there may be little or no time in which to consider this
decision. Even after the event, it may be very difficult to judge whether a
threatened person was really in mortal danger and had no other options. We
have the right to defend our lives, but it is extremely difficult to make the right
decision in the sudden and extremely pressurized situation when a person
thinks that a threat is being made on their life.

Even though we humans are genetically programmed to preserve our own
lives, some people do at times feel that they want to end their own life. The
drive for self-preservation is one of our most basic urges. This genetic instinct
has been, and continues to be, one of the primary factors in our personal and
evolutionary survival. This part of our nature has been enshrined in the rules
of most of our cultures and societies, so to want to end one's own life runs
contrary to both our personal inclinations and our cultural values. Yet a
minority of people do choose to end their own lives.

The human brain has different emotional states, ranging from happy to sad,
and from fulfilled and confident to depressed and pessimistic. All of these
emotions are part of the normal range of feelings that we humans experience
and the complex way in which our brain motivates and demotivates us. Our
emotions have an important influence on the decisions and choices we make.
It sometimes happens however, that some people get into a much deeper state
of mental depression than is part of our normal range of emotions. Of these
people, a further minority may feel that they want to end their life. This state
of mind may only last for some hours, or it may be experienced for a longer
period of time, but for most who reach this extraordinary state of depression
and hopelessness a time will come when the depression lifts and a personal

sense of purpose and motivation for life returns. A person who at some point feels suicidal usually only experiences this for a certain period of time. **Suicidal feelings are not felt during every minute of a person's life. This means that those of us who feel suicidal do have the capacity to feel normal and, given time, will do so again. So it is a mistake to end that extraordinary phenomenon that we call our life.**

We have to live our own lives and each of us has to make personal choices about those life options available to us. However, most people's lives are inextricably linked to other people's lives. Our choices also affect other people. Grief on the death of a loved one is a very painful experience. In the case of suicide, the normal feeling of grief is made worse because people agonize over whether they could have helped the suicidal person. Therefore, to end one's own life is not only one of the biggest mistakes we can make with our own life; it also deeply affects other people's lives. The attitude of The Natural Religion towards suicide is that not only is it a serious error to end one's own life (because suicidal feelings will change in time), but it is also very unfair to those people who know and love the person who ends their own life. It is not argued here that suicide should be classed as illegal in our laws, but society should instead give psychological counselling to help those who feel suicidal to see beyond those feelings.

Suicide is seen in many cultures as unacceptable and is in certain cases illegal. In other societies, suicide can at times be acceptable, even honourable, but even in these societies the circumstances in which suicide is acceptable tend to be very specific. Suicide that involves self-sacrifice in the interest of others or follows social expectations and norms can be acceptable in some cultures. However, particularly in the case of sacrificing one's own life in order to save the lives of others, the main purpose is not to end one's own life but to save life; the likely ending of one's own life as a result of trying to save others is an unfortunate side-effect, not the main aim. Whether sacrificing one's life in order to save others is an error or not depends completely on the circumstances. Suicide missions as an act of violent aggression or war are designed to kill people, not to save them, and these are errors. On the other hand, those people who sacrificed their own lives so that others could live are seen by many communities as heroes. If no other course is possible, then this may be the best solution in a bad situation. But, of course, it all depends on circumstances whether this really is the case. It is also the responsibility of all human beings to create the conditions in which such extreme and virtually insolvable problem situations do not occur. For a person to sacrifice their own life in order to save the lives of others as a complete last resort, with no other possible course of action available, means effectively that our human planning has failed.

To allow the self-sacrifice of a person to become so socially acceptable and honourable that it in itself becomes the reason why people end their lives is a corruption of an already undesirable and extreme situation. In this case, the reason why a person takes this extreme decision has now become social recognition, rather than to save lives, and this is a mistake.

People get sick, and some get so sick that recovery is impossible. In some communities, it is acceptable that a person's life is ended if there is no prospect of recovery and he or she is suffering greatly. The ending of a person's life when they are terminally ill can be their own wish or can occur without their permission if they are in a coma. Sometimes when a person in a coma is being kept alive on a life-support machine with no prospect of recovery, the decision is taken to turn off the machine, resulting in the death of the person. This is normally done in consultation with, and with the permission of, that person's family, since a person in a coma cannot convey their wishes to others. In some societies, it is also accepted that a person who is terminally ill and is suffering can themselves request that their life be painlessly terminated. This is illegal in many countries and many religions are also against this.

In both these situations a life is ended, and the time that a person has to live is shortened. One of the main aims of our existence is to prolong the time that we are alive. However, if a person is in a coma from which they will not recover, that person will never again consciously experience sentient human life. In the specific circumstances where there is absolutely no chance of recovery, and where the family and those who love this person agree to allow them to die, it is the best option and not necessarily a mistake. In the specific circumstances when a person is suffering severely and this condition will be fatal in the near future and there is absolutely no prospect of recovery, then for a person to choose to end their own life is most probably not an error.

The extremely bad situations, of an unrecoverable coma and an unrecoverable and soon to be fatal disease causing severe suffering, are extraordinary. In these extreme circumstances of a least-worse choice in an intractable bad situation, it is reiterated in The Natural Religion that the decision to allow the end of a human life or the choice to end one's own life does not in any way detract from the amazing, special and very precious nature of human life.

What is generally accepted as fair and unfair is not in all instances the same in every society. However, we all have personal needs, as well as requirements, which we expect from our communities. Like any living species, we have basic requirements to keep us alive and what is seen as fair tends to be those requirements. Basic needs for human beings are food and the resources to produce food. To take away a person's food source can be the same as killing that person. Hence the universal view that to protect possessions, particularly of food and where it comes from, is not only fair but essential. **To create a dependable food supply is probably the most essential task of our lives. We cannot plan anything else in our lives, including having and caring for children, if we are not reasonably sure of having enough food.** Therefore we have to protect the dependability of food and other resources with various types of ownership. Occupying a territory or food source is also at the basis of all other animal species' ecology. The specifics of how sources of food are protected vary according to the species involved, but dependability of food supply is an essential for all living beings. So also, human societies have to protect sources of food. To threaten these is an error and our social relationships cannot allow it.

Besides food, we also need other basic resources, such as shelter and protection. Our knowledge has allowed us to accumulate technology and we need to care for this technology and our other achievements so we can use it and benefit from it in the future. If technology is stolen or destroyed it can result in serious disadvantage to the people who lost it. It also means that the effort and resources it took to make this technology are wasted. We cannot afford to further develop and make technology if we cannot be sure that we will also benefit from it. We need to have resources that are dependable, and possessing them makes them dependable. Possessing resources in general is therefore an important aspect of our ecology. If we make choices that interfere with basic ownership and the dependability that it creates, then these are errors.

However, while every person needs the dependability of possessions, ownership has been used in the past to restrict the opportunity for other people to gain possessions. **Therefore, part of The Natural Religion is the concept that the prime function of ownership is to create dependability of resources, not to exclude people from them. Ownership should not be used as an excuse to withhold resources from those who need them.** The balance between sharing resources and ownership is examined in Chapter 8, which discusses the role of economics in our ecology.

We are a social species. Very few human beings do not live as part of a community. This means that we do not have total freedom to make all the choices we would like. This limitation of freedom is part of the greater overall limits within which the living system on Earth operates. All individual living beings need energy and resources, which other living beings also need. So, our social system and a certain amount of competition always limit our freedom to some extent.

The limitation of complete freedom, which people in a community inevitably impose on each other, is however compensated for by the advantages that we gain by being part of a community. Therefore if one does not contribute in any way to a community or one tries to be completely self-sufficient, this is normally a mistake.

Most of us live in communities and together we shape the community we are part of. We should therefore contribute to it in such a way as to try to create the type of community that we like living in. Whether our social environment and the type of life we live is comfortable, relatively stress free, caring, and allows us to maximize our life's potential; depends on how we shape our own society or community. If we leave it to others to influence and shape our community, then it may or may not be what we want. It is also unfair to let others spend their effort and energy in helping to form a community and not to contribute to it oneself. It is the position of The Natural Religion that it is fair that everyone in a community helps in some way towards shaping and operating that community. It is a mistake not to make an effort, together with other people, towards creating a community to our liking.

Like other living species, we compete for resources. This can give rise to conflict and people can feel the emotions of jealousy and hate. Despite the fact that these feelings may drive a person to compete more energetically, most people do not enjoy feeling these stressful emotions and try to avoid them. If we feel jealousy or hate and we allow these feelings to become important in our lives, we end up living in an unpleasant and stressful personal world. The objects of our jealousy or hate only affect our mental state through our own feelings. If we could control these unpleasant feelings, we would feel happier. It may be practical problems that give rise to these negative feelings, but it is better to deal with any possible problems when one is thinking clearly rather than when one is feeling powerful negative emotions. A fast and emotional response can, in fact, make the problem worse. **It is a mistake to allow feelings of jealousy, hate and anger to influence our choices and the way we experience and deal with life.**

Knowledge plays a crucial role in our lives. We receive it from other people either directly or indirectly and we in our turn also need to give knowledge to others. This is part of everyone's life. We also have the choice to give knowledge that is incorrect; in other words, we do not tell the truth all the time. Humans are not the only species that deceive others on purpose. Other animals also deceive in various ways in order to attain a goal, including a number of bird species, cats, dogs, several monkey species, as well as our closest genetic relatives: chimpanzees, gorillas, and orang-utans.

The reason why lying can deceive people is that most of the time the information we are given is correct. If the information we receive is more often incorrect than correct, then we would not believe it and a person who is lying would not succeed with the purpose of their lies. Lies and deceit only work if honesty is used for most of the time. For example with vervet monkeys, if one monkey tries to deceive the others too often with inappropriate alarm calls, the calls that individual utters will be ignored by the rest. Yet that same alarm call will be reacted to appropriately if uttered by more honest individuals. The same happens with humans – if people are known to lie, others will ignore what they say because they are known to be unreliable. It could be said that if we humans are honest most of the time and only deceive and lie very infrequently, then we could get away with it. However, we are very sensitive to the dishonesty of others because it could be harmful to us, so lies are often discovered. If a lie is discovered, people will remember this and any future information from that person will not be believed or will be ignored or, at the very least, carefully assessed and checked. The chances are great that even if we lie very infrequently, our credibility and standing in a community will be damaged. People who are known to tell the truth have a great advantage over people who are known to lie or who are suspected of lying.

Human ecology is very dependent on knowledge information and the honest and correct sharing and transfer of information is crucial to us. These are reasons why choosing to lie and deceive is deemed a mistake by The Natural Religion.

All through this book, we consider the question of a moral code or ethic based on reasoning, and the choices we make and how we should act are assessed based on the practical consequences of these actions. Traditionally, religions have advised, even imposed, a moral code because certain things are 'good' and others are 'bad or evil', often based on what is thought to be the will of supernatural gods and spirits. However, as we have seen, the approach as embodied in The Natural Religion is to search for the practical reasons for doing certain things and not others. We have arrived at a point in the development of human knowledge where we understand how life works, and the broader consequences of what we do, better than ever before. What is suggested in The Natural Religion is to use this understanding, in deciding which decisions we should take and to avoid making mistakes and errors.

For example, to recap, life is our most precious possession and we should fully appreciate our extraordinary good luck in being alive. Therefore, ending another person's life is a very serious error as is ending one's own life. In the intractable circumstances of a person suffering with an incurable, soon to be fatal disease, or of a person who has been in a coma and is being kept alive by a life-support machine for a considerable period of time and for whom there is no prospect of ever regaining consciousness, The Natural Religion accepts that there is moral justification in a person having the option to request their own death or, with the full permission of the family of the person in the coma, that he or she be allowed to die by turning off the life-support system. However, The Natural Religion strongly emphasizes that in both extreme cases (the final stages of a incurable terminal disease which is causing serious suffering, and a person in an incurable coma), every effort should be made to make sure that no cure or improvement of the very ill person is possible before ending that person's life is even contemplated. The Natural Religion suggests that strict civil legal criteria and specific practical procedures should be drawn up and followed in every case in order to eliminate errors and abuses associated with the ending of a person's life in these very exceptional circumstances. Again, The Natural Religion reiterates that our life is a most extraordinary, awe-inspiring and precious phenomenon and that it is only in these extreme and irretrievable situations that ending an ill person's life should ever be considered.

The ideas that form The Natural Religion describe a moral set of ideals, rather than legal rules that exactly prescribe what is allowed and is not allowed in specific circumstances. Any person should, in addition to forming their own moral opinions and adopting their own ideals, make sure that they do not transgress the specific laws of the land they live in or travel through.

The reality is that we humans make errors and mistakes, so how do we deal with that? We tend to be very sensitive to errors, particularly those made by others. It is important that we notice and correct mistakes so they are not repeated. However, once again, we need to strike a balance, this time between the extremes of an exclusively condemning and punitive reaction on the one hand, and an irresponsible disregard of errors on the other. **The first priority when addressing the issue of mistakes should be to try to ensure that they are not repeated. To achieve this, they should not be ignored, or only used for blaming, revenge and punishing. As a species we need to learn from our mistakes so we can avoid them in the future.** Depending on the type of error or mistake and the circumstances in which it occurred, it may be appropriate for a society to in some way punish the person who made the error. The primary reason why our societies punish errors is to attempt to stop them being made. Punishment is intended to encourage us all to stop and think and not to take what may initially seem an easy way out.

The approach suggested here is that we focus on our mistakes so we under-stand why and how they happened, so that if possible we can repair any damage that may be done by them, evaluate the effect of any punishment on their reoccurrence and learn not to repeat them in the future. The Natural Religion does not favour an exclusively disciplinary approach and takes the view that, while various forms of punishment are appropriate, the main aim of dealing with errors and mistakes should be to avoid them in the future. This would seem to be the best way to reduce unfairness and conflict in people's lives.

## 5.4: Being happy

Happiness is an emotion produced by the brain just like all other emotions such as sadness, excitement, boredom, satisfaction and frustration. These emotions are electro-biochemical activities in the brain. Our emotional state is like an automatic, often subconscious, evaluation of what is happening to us; we may not even be aware that we are interpreting particular events. These feelings well up inside us, rather than being the result of a conscious decision to feel like that. What decides the type of feeling or emotion towards a particular event in our life is usually the result of a combination of our genetic make-up and our experience of life so far. We also do our reasoned thinking with our brain, but we usually experience this as distinct from our feelings and emotions. Yet, our emotional state at any particular time strongly influences choices and decisions that we make using reasoned thinking. We try to make choices that will result in nice feelings rather than nasty ones.

People all around the world feel that the immediate reason for living life is to be happy. The feeling of happiness is probably the most wished-for human experience in the world. Happiness is therefore also the most important single reason for the decisions and choices that we make. The quality of human life is connected to the amount of happiness we feel.

Happiness is not only a human feeling. Many animals act similarly to the way humans do when we are happy. Examples include:

- jumping about with joy (e.g. dogs, mountain goats, horses and rhinoceroses);
- purring or grunting with satisfaction (e.g. cats, lions, cheetahs, bears, chimpanzees and gorillas);
- loudly calling with excitement (e.g. elephants, dogs and chimpanzees);
- sliding down slopes (e.g. otters, bears and ravens);
- swimming around in circles and performing water acrobatics, (e.g. beavers and otters) and jumping into the air (e.g. dolphins and whales);
- jumping into the water, climbing out and jumping back in again (e.g. dogs, otters and hyenas).

Many of these types of behaviour are reported by those who observe them as carried out with a joyful exuberance, body language and expression that we interpret as signs of happiness. It does seem that the feeling of happiness and other feelings that we have are also produced by the brains of other animals and like them, humans developed these feelings in our evolutionary past. So, these feelings are assumed to be of evolutionary advantage rather than just a by-product of the way our brain operates.

The question is often asked, 'Why don't we feel happy all the time?' **Most people want to feel happy all the time, yet it seems to be impossible for a person's brain to produce the feeling of happiness continuously.** It is a well-known phenomenon that things that initially make a person very happy gradually lose this effect. It is as if we cannot feel happiness unless we also regularly feel other feelings. This does, however, mean that we keep trying to feel happy. We may attain happiness, but it only lasts for a certain period of time, after which we feel other feelings and we then try to achieve happiness all over again. The result is that striving to feel happy, keeps us active all the time and perhaps this is the way in which happiness motivates us to look after our interests energetically. Efforts to achieve happiness range from creating comfortable and safe physical surroundings, to trying to influence and control our minds directly to make us feel happy. Religions are amongst our social organizations that suggest ideas and choices which will result in us feeling happy.

Although it may feel to us that the reason why we act in a certain way is to feel such nice emotions as happiness, love, satisfaction, control, contentment, comfort, security and tranquillity, this, in a sense, is probably putting the cart before the horse. The actual situation, and view included in The Natural Religion, is that what we achieve by the way we act are the real reasons. Nice emotions have developed so that we are more inclined to engage in these beneficial activities. **Nice feelings function as a type of reward for doing things that tend to be good for ourselves and our communities. Put in yet another way, evolution has developed feelings in us that drive us to act in evolutionary advantageous ways.**

People all over the world are different. We have different genes and everyone has different life experiences. It is impossible for two people to have had exactly the same life experience in every respect. Even identical twins who have the same genes and are reared in the same family, do not have identical experiences and are therefore slightly different. For these reasons there will always be slight differences in what makes individual people happy. However, as a species, humans all have the same basic needs and we share many of these needs with other animals.

Food is probably our most basic need, after breathing air and drinking water. When we are hungry and need food, getting and eating food and even the smell of food, is pleasurable. However, once we have eaten, food does not have the same appeal anymore because when we have eaten, we need to stop eating and get on with digesting our food. To continue eating when we should be digesting is not good for us. The feeling of having eaten too much can be very unpleasant, even sickening. In this situation the mere smell of food can have the opposite effect to what it has when we are hungry. Eating too much is not good for us personally, both immediately and for our future health, and it is also a mistake in a broader context because the food one eats when overeating is wasted and could have been used by people who are hungry.

It is important to every individual person to be safe and have shelter. We also need to be safe from diseases; worldwide over the course of human history, diseases are the greatest cause of death. We also need to be safe from physical accidents caused by our own errors (such as road traffic accidents), as well as external factors such as the weather, dangerous terrain and other animals. In addition, we need to be safe from attack by other people; humans do attack each other for various reasons, often directly or indirectly related to competition for resources. So, when we strive to feel comfort and security, we are effectively putting ourselves in a situation where we minimize the chances of harm through diseases, accidents or attacks.

A species does not survive if it does not reproduce. Any species that has survived has to have a strong urge to reproduce. Humans too have this urge. Everything that is part of our reproductive process has strong feelings of happiness and emotions associated with it. We dedicate a lot of time and effort to meeting and choosing a prospective sexual partner, forming relationships, having sex, nurturing and rearing children, and helping them to be able to have children in their turn.

Besides food there are many resources that we humans need. Implements and technology (such as clothes, houses, tools, medicines and knowledge itself) all make our lives easier, safer and more predictable. We are able to achieve more using such resources and they give us feelings of security and control.

In most cultures, playing, art, sport and other leisure activities are a great source of happiness and amusement. These activities themselves may not seem to be of great practical value. Yet, if resources allow, most people enjoy them and therefore engage in them. The exhilaration of physical exercise, which our bodies need, and the mental stimulation and relaxation we get from the many different forms of art give us enjoyment and happiness.

Playing generally consists of some activity that is not serious. It allows our brain and body to relax and also to practise things without the result being important. In real life, results matter, so a certain amount of stress is associated with them. If we make a mistake in play, it usually doesn't really matter. Play can be a relatively stress-free way in which to practise activities and make mistakes, which in a real-life setting could be serious. As well as ways of inventing and creating new things and ideas, playing and art are usually accepted as being leisure activities for enjoyment and are not expected to produce immediate practical achievements and results, so they allow much greater freedom to experiment. This freedom can lead to unexpected results. In fact, many a ground-breaking discovery was made by accident. Therefore, playing, art and sports actually do have a number of practical functions, although this is not immediately obvious and it tends not to be the reason why people decide to engage in them. Most people engage in these activities just because it amuses them and makes them feel happy.

Social animals live close together and help each other in various ways. We feel drawn to each other and need the stimulation of other people. To withhold all human contact from a person for an extended period of time is a form of punishment. Contact with other people, even of a non-friendly nature, may even be better than never having any contact with anyone. We are pro-grammed to need a certain amount of human interaction. We enjoy and relax in the company of friends; just chatting with other people is something many of us enjoy, even though the topic of conversation may not be of immediate importance. The social contact alone is enough to make us seek the company of others.

Our love of those we have close relationships with, enjoyment with friends, and the satisfaction of achieving goals as part of a group – all cause us to seek the company of other people. Feelings of love, company and acceptance make us strive to be part of social systems with our fellow humans.

Most people want to be happy and also know that others want to be happy. Real happiness is not all that easy to attain and, as mentioned before, when it is attained, it is only felt for a certain period of time. Because happiness is much sought after and valued, if one person can give another person happi-ness, then this is something that is greatly treasured. To help another person to feel happy is one of the greatest gifts one person can give another and is therefore a major achievement. Since we are a social species, we have the ability to affect the feelings of other, and to affect another person's feelings also affects our own feelings. We feel a sense of satisfaction about having accom-plished such a powerful thing as making a person happy. To make someone else happy also makes us feel happy. Helping another person is part of the framework of our social system, and helping and making someone happy is part of our close relationships. However, helping and making someone happy with whom we don't have a close, or any, relationship can also feel very satisfying. The happy surprise of a person who does not expect it, can be very rewarding.

**When we feel depressed, dissatisfied or bored, it can be difficult to make ourselves feel happy and content again. One of the best ways to try to feel happy again is to ignore one's own feelings by, in a manner of speaking, putting them on hold and instead concentrate on helping others and making them feel happy.** Even if we are only partially successful in this, we will have diverted our attention from ourselves and will gain some feeling of satisfaction, perhaps even happiness.

Why do we feel happy when we make other people feel happy? As social animals, we are very easily affected by the moods and feelings of those around us. This is certainly the case when we are with people we know, but can also be the case in a crowd of people we don't know. When those around us are happy, it tends to be infectious and we are also inclined to become happy. To give this feeling to another person can result in a real sense of accomplishment. We are programmed to feel happy when we make others feel happy and this is part of our social system. The reward of our own happiness encourages us to help others.

One of The Natural Religion's main aims is to help people achieve the best quality of life possible. This includes trying to maximize the feeling of happiness that we feel. Keeping oneself and one's family alive and safe, being part of relationships, having and caring for children, being part of and contributing to one's community, making other people happy, and engaging in relaxing and amusing activities – all give us various types of happy, loving, secure, satisfied and contented feelings. Nevertheless, our feelings of happiness only last for a certain length of time. Changes in circumstances and the way our brain functions mean that no activity or situation can result in us feeling continuously happy. Unlimited food, a loving relationship, limitless wealth, dedicating oneself completely to leisure activities, even being perfectly healthy, do not result in a person being happy all the time.

The new approach to religion offered here, like other religions and organizations, cannot guarantee happiness for every moment of an entire life. Therefore, while **The Natural Religion suggests that an active but balanced approach to life will most likely maximize our feelings of happiness, it is not realistic to expect to feel happy all the time.** Helping others and making them happy is a particularly reliable way to create feelings of happiness and satisfaction in oneself. Not everyone is in a position to attain all of life's main goals, so the extent of our feelings of happiness, satisfaction, and contentment is also influenced by what our expectations are of life. Therefore, **the advice included in The Natural Religion is that the most likely way to feel happy** is to actively care for our own and other people's well-being, and to strive energetically to attain realistic goals and expectations in our lives. One life expectation of most people is to have children, and therefore the following chapter is about the important and multifaceted issue of sex and our reproduction.

# Chapter 6

# Sex

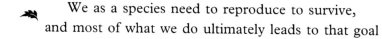

We as a species need to reproduce to survive,
and most of what we do ultimately leads to that goal

## 6.1: Our main task

Because sex and reproduction are so important to us, because the actual practical details have such far-reaching consequences on the course of our lives, and because people are at times embarrassed about discussing this in depth; this chapter goes into the emotional and biological specifics of our sex lives. Our species' existence depends on reproduction. The process of reproduction goes back through thousands of generations to the emergence of our species, and beyond that through hundreds of millions of years, and all our evolutionary ancestors, to the original single-celled organisms that populated the world. Reproduction forms the centre of our human existence. Most of our activities are either directly or indirectly related to our reproductive process. It is ingrained in our psyche, it is the ultimate aim of many of our thoughts and feelings. Creating our descendants and how well we prepare their environment for them, is the measure of our success as individuals and as a species.

The basic understanding of The Natural Religion is that preparing for reproduction and creating and caring for human life, is human beings' main task. However, contributing to the reproduction of our species does not solely mean having genetic children, it also means helping to shape and maintain the circumstances which future generations need to live and reproduce. **So, not all adults necessarily need to have children for our species to continue to exist, but all of us do need to help with ensuring our survival.**

## 6.2: Two sexes

The human species like most other living species on Earth is made up of two sexes. In order to produce a new human being, a man and a woman both have to contribute to the single living cell from which a child develops, this is called an ovum. Women provide the egg, which is the largest part of the ovum cell. Men only contribute genetic material for inside the nucleus of the ovum, which is contained in the male sperm cell that consists of a small head with genetic material in it, and a tail that the sperm uses to swim. The sperm, which is much smaller than the egg, has to be given by the man to the woman and placed close to where the woman has the egg so that the sperm can swim toward the egg. The process by which a man gives millions of sperm to a woman and puts them close to her egg is having sex.

One of the millions of sperm arriving at the egg touches it with its head and bores through the outer wall of the egg, and then injects the genetic material from its head into the egg. The sperm's genetic material goes to the nucleus of the egg, and there it combines with the egg's genetic material so that the egg then becomes a fertilized ovum.

In humans, the fertilized ovum cell stays inside the mother and starts to grow. It first divides into two cells, then four, then eight, then sixteen and so on. Over a period of nine months the fertilized ovum grows into a baby and then the mother gives birth. Although a baby is able to live outside the mother after its birth, it is completely helpless and needs total care; and everything it needs like food and protection has to be provided for it. As the baby develops into a child, it continues to depend on adults for most of its requirements. A person is only fully adult when he or she is able to provide for his or her own needs. With human beings this may take more than twenty years. All during these years, the child's parents and/or other adults need to care for it and provide it with resources and protection.

One question that can be asked is: why does the cell which forms the beginning of a person need a contribution from both a woman and a man? Why can't a woman just produce an egg which can develop into a baby without needing genetic material from a man? In fact, the females of a number of animal species can actually do this, stick insects are one example; they produce eggs which develop into adults without receiving genetic material from a male. This type of reproduction is called parthenogenesis.

We humans have a double set of chromosomes in the nucleus of our cells. Chromosomes are the large strings of molecules on which our genes are encoded. When both the egg and sperm develop, this double set of chromosomes undergoes a process of doubling and splitting, at the end of which both the egg and sperm end up with a single set – half the normal number of chromosomes – in a process called meiosis. These two single sets join together when the sperm fertilizes the ovum, giving the child a new double set of chromosomes made up of one set from the mother and one set from the father, so the child will not be exactly like either the mother or the father. In addition to this, when the eggs and sperm are formed in the parent during meiosis, the chromosomes do not divide in the same way every time. If they did, each child of a particular man and woman would be genetically the same. This is not the case, because when the sets of chromosomes split during the development of sperm and eggs, each chromosome of a pair are distributed at random between the resultant sperm and egg cells. In addition, genes swap over between the chromosomes during the process of meiosis and not in the same way every time. The result of this process is that egg cells as well as the sperm cells all end up with different combinations of genes on their single set of chromosomes. This is why children are not exactly the same as their parents, and children from the same parents also differ from each other.

Why do we have such a complicated process of reproduction? Why did life develop so it could produce offspring which is different from itself? The reason for this genetic diversity is that, as we have seen in previous chapters, the environment on Earth changes all the time, and if each generation were to be the same as the one before it, then life would end up not being adapted enough to survive as our Earth's environment changes. With genetic diversity, at least some individuals will be able to deal with environmental changes, as long as these changes are not too extreme. Those who are better able to survive will produce more offspring, and so a species changes to stay adapted to its circumstances. All living biological species have to change in order that they can continue to survive in a constantly changing earthly environment; this is the way life works, this is evolution by natural selection.

In species such as stick insects where the females parthenogenetically produce eggs which don't need fertilization to develop, the daughters are all genetic copies of their mother. These offspring are the same as the parent, and so do not change from one generation to the next. However, interestingly, such species are able to change to sexual reproduction periodically when environmental conditions change and demand it. So these species as a whole still have the capability to evolve and adapt to changing circumstances.

Therefore, the reason why we have two sexes amongst humans is to be able to mix up our human genes and bring them together in different combinations in individual people. The ecology of life, including our human ecology, has forced us to evolve like this so at least some of us are prepared for when our environment changes. This, in combination with different life experiences, is the reason included in The Natural Religion why no two people are exactly the same.

## 6.3: Reproducing our brain

All through this book we keep coming back to the huge importance of knowledge information in our lives, and how it distinguishes us from other animals species. As we saw in Chapter 4, prehuman fossil-finds indicate that brain size increased as our evolutionary ancestors evolved. As also mentioned before, the brain size of a baby chimpanzee is approximately half that of the adult chimpanzee, while that of a human baby is only between one third and a quarter the size of a human adult's brain. A human baby is much less developed and less capable at birth than a chimpanzee baby. For a human baby to reach the level of development of a baby chimpanzee at birth, human pregnancy would have to far exceed nine months. We know that human babies are helpless and they are completely dependent on care from adults. All during the time that we develop from babies into infants, children, and teenagers; we need care, protection, and education. Gradually we become more able to care for ourselves but it may take more than 20 years before we are fully capable of doing this. The slow development of human beings is primarily attributed to our brain's small size at birth, and to the time it takes for the brain to mature. As babies and children we need physical stimulation for our bodies to develop properly, as well as mental stimulation for our brain to develop properly. Prolonged care for our children is an integral part of creating our capacity for knowledge, and therefore crucial to our human ecology.

A basic principle of The Natural Religion is, therefore, that good and prolonged childcare is an integral part of our knowledge-dominated ecology and is essential in order that each person can develop their potential and lead a full adult life.

## 6.4: The role of human sex

When we wish for and anticipate sex, we become excited and our body prepares itself so that it can carry out the sexual act. In a man the penis enlarges and stiffens due to engorgement with blood, the testes introduce sperm into the tubes which run from the testes to the penis and seminal fluids are added to the sperm as they travel along these tubes. In a woman, mucus is produces in the vagina and the outer female genitalia become somewhat swollen. Because the vagina is lubricated by mucus and fluids, the stiff penis can be pushed into the vagina and sexual excitement in both partners can give great feeling of pleasure and well-being. Both the man's and the woman's genitalia become sensitive as part of the sexually excited state of the body, and rubbing them together and pushing the penis in and out of the vagina is then a very pleasurable experience.

After some time of sexual stimulation one or both partners can reach a sexual climax or orgasm. In the man this means that he squirts a mix of sperm and seminal fluid out of his penis into the woman's vagina. The end of the vagina at this stage will have opened up, resembling a tiny cave. At the end of the vagina there is the cervix which forms the entrance to womb. When the woman reaches a certain point in sexual excitement, the cervix elongates forming a small tube and hangs downwards into the little cave at the end of the vagina. The semen has now collected in a puddle on the bottom of the little cave and the end of the cervix hangs into it. A few of the sperm in the seminal fluid use their tails to swim into the hole in the end of the cervix, and swim up the cervix and into the womb; the cervix then retracts again cutting off access to the womb. The sperm travel through the womb into the fallopian tubes which lead to the woman's ovaries, and there one of them fertilizes the egg if an egg is present. The eggs are produced by the ovaries and travel down the fallopian tubes. Normally most of the semen stays behind in the puddle at the end of the vagina, and is ejected again out of the vagina a short time afterwards.

All during sex a number of things are happening. One of these can be that the woman feels an orgasm, this includes a rippling of the muscle in the walls of the vagina. Orgasms in both women and men give an extremely pleasurable surge of feeling, part of which is muscular spasms in many other parts of the body besides the genitalia. People rank the gratification and enjoyment of sex as amongst the most pleasurable experiences of our lives. Evolution has made sex a great physical pleasure, and it is one of the most desired feelings in human cultures around the world.

Other animal species also experience the sex drive. In most species, the males tend to be ready to have sex for longer time periods than the females, in fact, females tend to only be receptive for sex when they are fertile. Some species reproduce once per year, and then the female is only sexually receptive at that time. Other female mammals may only want to have sex once every few years, if the pregnancy and rearing of the young takes several years.

On the other hand, in humans women are receptive for sex all the time, which is unusual amongst mammals. In fact, a woman's fertility and menstrual cycle is such that it is often not exactly clear when she is fertile, which is also unusual in the animal world. This obscuring or keeping secret of a woman's reproductive cycle is called 'sexual crypsis'. In contrast, most female animals give very clear signals, either visually or by smell, when they are fertile. This means that time and energy is only devoted to sex for the limited period of time when females are fertile and not when they are infertile, which in most cases is for the majority of the time.

Because of sexual crypsis we humans tend not to be aware when exactly a woman is fertile or infertile, so we have sex regularly with the result that most of the time there are sperm present in the woman's fallopian tubes ready to fertilize an egg, when her ovaries produce one. Both women as well as men enjoy and devote time and energy to sex at all stages of women's fertility cycle. Since sperm remain viable in a woman's fallopian tubes for up to five days, live sperm can be present all the time.

Courtship and sex are intense activities and take a lot of energy, during this time animals often don't feed and therefore weaken. Added to this, when an animal's attention is occupied with courtship and sex, it is not as alert to the dangers of predators. For these reasons it is more usual that animals do not engage in sex if the female is infertile and reproduction is not possible. We humans, on the other hand, engage in sex while women are infertile, or are fertile but do not want to have a baby. We devote a lot of time, energy, and resources to courtship and sex, even though the chances of a sperm fertilizing an egg are small or non-existent. So why do we humans have so much sex, and why is so much of the sex that we engage in unlikely to result in pregnancy?

**Reproduction in humans is not just a matter of delivering the sperm close to a fertile egg, nourishing the fertilized egg and foetus to full term in the womb, and then giving birth to the baby. Until a baby is adult and itself in a position to have children, the reproductive process is not complete.** Small children bring out protective and nurturing emotions in people. Their helplessness fills adults with feelings of tenderness and love. Children who are genetically related to us are of special importance to us, as they represent the continuance of our own particular set of genes; they and their children constitute our personal evolutionary success, our genetic legacy. Therefore, it is in the personal genetic interest of both the mother and the father to care for them. Two people are better able to look after children than one person, and two people are also likely to have more resources to give to children than one person. The better a child is cared for, the more likely it is to reproduce, and the greater the chance that its genes will be included in future generations. Genes from people who care for their children will have a better chance of getting into future generations than genes from people who don't.

There is therefore an evolutionary reason why a man or a woman should form a relationship with the person with whom they reproduce, which lasts far beyond the time it takes to fertilize an egg and for the baby to be born.

**So, human sex is not necessarily only for fertilization of an egg. It has become part of the way in which a relationship is formed and maintained by helping to form a close and intimate emotional bond between a woman and a man.** Men and women are sexually attracted to each other, even when they don't know each other. Just the visual impression of a person can be sexually attractive to us. As a woman and a man get to know each other's personality, the attraction between them changes to a more emotionally dependent one. It is at this point that a strong personal relationship can start to grow. The very intimate and personal nature of sex, as well as the excitement and pleasure of having sex, can all be part of the close emotional bond between two people. The strong feeling and attraction which a woman and a man can have for each other is being in love and their relationship a romantic relationship.

Intimate sex is an important part of romantic relationships, and relationships bond people together so they can depend on each other more. Sex helps to form a close relationship between a woman and a man so they are more likely to help each other raise their children. Because relationships are so important to the success of us as a species, the need to have relationships is deeply ingrained in us. We have evolved an emotional dependency on relationships, both as children and adults. Most relationships in our lives are not romantic ones, but sex is normally a part of those that are.

Humans are not the only animals in which sex has taken on a broader role than purely to ensure that a sperm will fertilize an egg. Our closest living genetic relative pygmy chimpanzees, also known as bonobos, use sex and rubbing their genitalia together as part of normal everyday social interaction. Begging for food, greeting, or appeasing each other after conflict are all frequently accompanied with genital contact of one form or another. Bonobos use sex even more widely and more often than we humans do, including between individuals who are the same sex and could never reproduce. The only individuals who tend not to use sex as part of social exchanges are mother bonobos and their adult sons.

A minority of other animals, such as some bird species, lions and common chimpanzees also engage in sex more often than is strictly necessary for fertilization. Judging from the reactions of certain other animals species it would seem likely that, like ourselves, females as well as males experience physical pleasure from having sex. However, with most other animals, sex is only associated with fertilizing the eggs. Other social interactions, including cooperation in caring for the young, are generally not associated with ongoing sexual behaviour.

So, the function of sex in humans has broadened from fertilization of the eggs to facilitating care for children by establishing and maintaining close relationships between mother and father. Sexual crypsis is part of this because it ensures that we need to have sex comparatively often for fertilization to take place. The following list shows how sexual crypsis and relationships are part of our overall life and ecology.

- Sexual crypsis leads to having sex often.
- Frequent sex facilitates long term stable relationships.
- Stable long-term relationships of the carers promotes good childcare, development, and learning.
- Our long childhood and adolescence allow for the physical development of our large brain and learning knowledge.
- Our large brain and its power of knowledge has made human ecology the extraordinary success that it has been.

Of course, human sex is not limited to the relationships which care for children. It can also be an important part of close personal relationships that don't have children, as well as short dalliances in which sex is the only reason for meeting.

From the broader role that sex plays amongst ourselves as well as some other animals, the following conclusions are drawn and included in The Natural Religion. The way that women and men cooperate and combine in placing sperm close to the egg is normal and natural in every way and is not in any way impure, indecent, dirty, or obscene. The fact that human sex is an intimate activity which is usually carried out in private does not mean that it is in any way shameful or degrading. The apparently involuntary wave of passion which accompanies sexual excitement is not animalistic in any derogatory sense. Sexual passion is part of our normal range of sensations and emotions, like sneezing or coughing, which other animals of course also experience. Human sex has come to serve a broader function than strictly placing millions of sperm close to an egg so that one sperm can fertilize it. The Natural Religion sees the intimate nature of human sex as part of forming and maintaining close emotional relationships which creates environments conducive to good childcare and education. However, people having sex solely for the pleasure of it, and with no intention of forming a relationship, also happens.

## 6.5: Family relationships

We humans are a social species, and the most basic social unit in our ecology is the family unit. A family usually consists of individuals that live together and give each other help and support, who may be emotionally attached and more often than not are also genetically related to each other. One of the most important functions of the family is to care for children. Families can take many different forms. Mother, father and children, together with grandparents and sometimes also aunts and uncles, each with their children, can make up a relatively large extended family. A nuclear family is made up of just a mother and father and their children and some families have just one parent with children. Aunts, uncles, or grandparents often also care for their young relatives, but children are typically looked after by one or both of their genetic parents. But, circumstances such as death of one or both of the genetic parents or a lack of resources can result in children being cared for either in another nuclear family or larger social units by people who are not necessarily genetically related to them.

The lives of members of a family are usually closely intertwined, people within a family often depend on each other for practical help and emotional support. The adults who look after the children depend on each other for assistance, and children depend on the parenting they receive. Much depends on the relationships within a family; the way members of a family relate to each other determines the atmosphere in a family home. For a child its home-life is what it most closely personally identifies with. The character of a family forms the background to, and influences, people's lives long after they are adult and no longer live as part of that family. Because the relationships within a family are so close and we depend on them so much, our expectations of our family life are generally greater and more demanding than our expectations of life outside the family. Based on what we know of the role of families in cultures all around the world, The Natural Religion concludes that nuclear or extended families normally provide the best environment in which to care for children.

When a man and a woman meet and are attracted to each other, they will each have expectations and hopes for their future life, but the chances that these expectations will be exactly the same are low. There are a number of reasons why women and men are likely to have differing expectations. Just the biological differences between the two sexes means that men and women's expectations differ, but more of that later. Any two people will not have exactly the same personality, nor will their experience and background be identical, and these are also reasons why their hopes and expectations will differ.

When two people are attracted to each other they tell each other their expectations of life, this is one of the functions of courtship. The courting couple will try to judge whether they are compatible, and whether and to what extent they may have to compromise in order to have a relationship. They will also try to find out to what extent their partner is willing to help them achieve their hopes in life. Most people will be prepared to make compromises and help their partner to achieve their aims in order to have a relationship. However, the advantage of having a relationship must be greater than the disadvantage of compromising. Helping one's partner achieve his or her hopes in life is particularly worthwhile if they also help you achieve yours. A man and woman can achieve more as part of a good personal relationship than they can on their own. If, on the other hand, a person is compromising much more than they are benefitting and that person's expectations of the relationship are not met, the relationship will be an unhappy one and it might even end.

Returning to the biological differences between women and men, one of the fundamental reasons for having a relationship between a woman and a man is to have children. The act of having consensual sex in order to fertilize an egg and give life to a new human being can only be carried out if a woman and a man both contribute and cooperate. Even just having sex with another person involves compromises, because it can result in contracting a sexually transmitted disease. To trust someone not to infect you with a disease is taking a chance and, in fact, is making a compromise.

The prospective mother is also compromising her health to a certain degree during pregnancy and giving birth, as not being pregnant carries less health risks than being pregnant and giving birth. In addition, the mother devotes energy and resources to caring for her child, this also compromises her self-interest as she would personally be better off if she used that energy and resources for herself.

The father cannot give birth to a baby, but he can provide resources and help to the mother during pregnancy and to the mother and the child after the birth. Like the mother, he compromises his personal well-being by giving his resources and energy to the mother and child. The reason we developed relationships is that they allow us to achieve things which are either more difficult or we cannot achieve on our own. Fertilization of an egg we cannot achieve on our own and caring for children is more difficult on our own. The Natural Religion points to the advantages of a relationship such as trust, help and loyalty, which have the potential to allow the partners to achieve more together than the sum of what they could achieve on their own.

On a very basic biological level, both a woman and a man may want certain things which are not in the interest of their partner. However, in order to have a relationship, a man and a woman have to reach a compromise or balance; and as with so many balances between two extremes in the ecology of life, the extremes are usually not stable.

At one extreme a woman may want to have sex and children with a more attractive, dynamic, and also more permissive but less dependable man, than her permanent partner is. If a woman has children with a permissive very sexually active man, her children are also likely to be more permissive as adults than if they have a more dependable and less permissive genetic father. This means that the mother's own genes are also passed on to more people by being linked in her children to the permissive genes of their father. So, it is in the genetic interest of the mother to have particularly sons by a permissive father, as they potentially can have many more children than her daughters. Her own genes are effectively 'piggy-backing' on the permissive genes her children received from their father and being reproduced more often, because being permissive they have sex with more partners. At the same time, she may want to keep her permanent partner who has resources and is more dependable to help her care for her children, but some or all of her children need not be his children. In this way the mother's genes are more likely to pass on to future generations with a more permissive genetic father, and also benefit from the support of a less permissive but more dependable nurturing male parent with resources.

At the other extreme, a man may want to have one steady partner who does not have sex with any other man. In this way, he is sure that the children he is helping to raise, and in which he is investing his time and resources, are really his genetic children. In addition he may also want to have sex with as many other women as possible whom he does not know well, and if these women have children he is not sure whether they are his or not because his relationships with these women are not close. But, he is not very concerned about this as he gives little help towards caring for these children. It is not in his genetic interest to devote his energy and resources to children if he is not sure whether he is their genetic father or not. Yet, if he has sex with many women he could potentially pass his genes on to many children.

For both the woman and the man it may make basic genetic sense to have sex with others besides their permanent partner, but it is not in the interest of their partner. It does not make genetic sense for a man to devote his time and resources to another man's children, nor is it in the interest of the woman if another woman lays claim to some of her partner's time and resources which could otherwise be devoted to herself and her children.

For both partners there is also the threat of sexually transmitted diseases if their partner also has sex with other people. Sexually transmitted diseases have been part of human ecology for a very long time. The more sexual partners we have had, the more likely it is that we will be infected. Both the woman and the man run an increased risk of being infected by their partner, if their partner also has sex with other people. The only sure way not to run any risk of infection is not to have any sexual partners at all, but the human species would not last long if everyone did that. Hence, the safest way of having sex is to have sex with just one partner who does not have any diseases, and who is dependable and can be relied upon not to have sex with other people. So, sexually transmitted diseases are another reason why it is not in the interest of people in a relationship to accept that their partner has sex with others.

A relationship based on the male or the female extremes is normally unstable because the two partners will most likely not accept each other's extreme, because they would not benefit from the relationship and so the extremes will be unstable. **However, if both partners compromise and a balance is reached between the two extremes in which neither engage in sex with other people and both cooperate to create and care for their common children, a relationship can exist and be stable.**

The extreme ideals for the children of a relationship are also not exactly those of the mother or the father. Children from the point that they are born act in such a way as to attract as much attention, help, and resources from the parents to themselves as possible. This can include resources which the parents may need for themselves or for their other children. A child may want to keep the full care of its parents all for itself, but if the parents have sufficient resources it is in their genetic interest to have more children. So the children also reach a compromise with their parents, because they are completely dependent on them and therefore they have to make it work.

Apart from the reasons already mentioned why both men and women should want to stop their partner having sex with others, there is another reason why having sex with others besides one's partner has problems associated with it, which is to do with relationships. We have evolved the capacity to form very strong and close relationships with our sexual partners. As sexual partners tend not to be closely related, they often don't know each other before they are attracted. Yet these relative strangers can in a short time fall in love and form a very close romantic relationship.

This sudden surge of feelings that we have for a person we are attracted to can be an overwhelming experience. It gives a sense of euphoria which does not seem under conscious control to the person who is feeling it, and gives us a strong urge to be with that person as much as possible. We cannot imagine that we will ever feel any other way. This intense feeling of love for a person has been celebrated in all the different art forms. **Love is claimed to be one**

**of the most exciting experiences of a person's life, it can be so strong that people have even braved the threat of death to be with the person they love.** This phenomenon of sudden very strong emotional attachment to a person forms the beginning of a relationship between two potential sexual partners. **Love helps to create a strong relationship in a short time.**

As we saw previously, a strong relationship between sexual partners is very important for the success of their children. People who love each other, want to help each other. The love and emotional strength of a relationship gives a couple the courage to trust in each other. The strong emotional bond between sexual partners makes it more likely that the partners will not try to attain their own extreme genetic interest, but compromise with the interest of their partner and this is seen as a sign of love for each other. **The Natural Religion warns of three basic problem of having sex with other people besides our permanent partner. It can bring genetic parentage and willingness to devote resources to childcare into question, it poses an increased health threat from sexually transmitted diseases, and it threatens the trust which exist in a relationship and thereby also its stability.**

Our human relationships have a great influence on our personal well-being, our emotional welfare depends to a great extent on them. Relationships can be very strong, such as those within a family, which can be so strong that they often survive quite heated arguments and disagreements. At times we don't fully appreciate how much a relationship means to us until we feel grief when it no longer exists. We have been genetically programmed to have these feelings of emotional attachment and it is these feelings which help to keep a family together.

In the case of sexual partners the initial strong feelings of love and sexual excitement for each other normally change after a period of time. If the relationship grows, these feelings may develop more in the direction of those that relatives in a family have for each other. Yet the sexual attraction can still continue to be part of the relationship. Why do these changes take place?

As a couple meet and begin to know each other they need to form a strong relationship rather quickly, this is an emotionally exciting time. Establishment of a relationship is important so people can get to know each other, allowing each partner to try to judge whether the other is dependable and reliable. They also have to judge whether their expectations are compatible, and to what extent they will have to compromise in order to have a relationship.

A couple typically wishes to be sexually active at this stage, and this promotes intimacy between them and knowledge of each other. However, actually conceiving a baby at this early stage is often not intended as the partners may feel that they do not know each other well enough yet to want to have children together. The couple may choose their particular sexual activities so as to avoid conception for instance by being sexual intimate but not having full sex, or by practising contraception. Family planning is dealt with in more detail later on in this chapter. Once a couple feel that they know each other well enough and love each other, they may commit themselves to their relationship for the longer term. It is at this stage that they may want to have children when resources and circumstances allow. The function of sex then also includes fertilization, as well as an intimate activity which the couple engages in for pleasure.

When the couple have the number of children they want, they will want their sex not to result in a baby, and while they may still engage in sexual activity for pleasure and pair bonding, they will try to avoid conception. At this stage their sexual activity may also not be as frequent or intense as earlier in their relationship. The relationship itself has had time to develop and grow and may not depend as much on the intimacy of sex as it did before. Sex for pair-bonding and enjoyment has done its work and now that the relationship is established, sex may only be for pleasure and maintaining the relationship. **As a relationship develops further and the emotional bond becomes firmly established, sex may gradually become a less significant part of the relationship. If the couple still love each other and feel that their expectations of life have been sufficiently fulfilled, their relationship can be deeper and stronger than ever.**

Of all the relationships which people have with each other, most relationships are actually non-sexual. In a family of two parents and two children, it is only the relationship between the parents which is a sexual one, the other five relationships, between each parent and each child and between the two children, are all non-sexual. A relationship which started as a sexual one can progress to the stage that it is effectively a non-sexual family relationship.

Love and relationships are not only important in people's lives, but are also important to the success of humans as a species. People's emotions have evolved in such a way that we depend on relationships. This emotional dependence of people on each other is so strong and ingrained in our psyche that we have many relationships which are not for reproduction, don't involve sex, or whose primary function is not to directly help each other. The primary reason for many relationships is that we simply enjoy the other person's company, and these can exist both within and outside of families. The Natural Religion points out that while relationships are central to human reproduction, many sexual partners have strong relationships, but do not have any children. This is not unusual, most populations of animal species consist of a certain percentage of individuals who for various reasons do not reproduce. The same is true for human populations.

Because sexual partners are generally not genetically closely related, in most cultures and societies their relationship has been given formal recognition and is officially made into a family by means of a marriage, and gives the partners the status of husband and wife. The most common type of marriage is a monogamous marriage where one husband and one wife live together as a nuclear family or as part of a larger extended family, and care for their children.

There are also polygamous marriages in which one husband has several wives or one wife has several husbands. If one husband has several wives it is a polygynous marriage, and if a wife has several husbands it is a polyandrous marriage. While most marriages around the world are monogamous, there are more societies and cultures which allow polygamous marriages than those which only allow monogamous marriages. But, in most societies which allow polygamy, the majority of marriages are still monogamous. Most wives prefer not to share their husband with other women, however if a significant number of men are so poor that they can hardly keep themselves alive, they then cannot support a wife and children. This means that an equal number of women do not have husbands, and for these women it may be better to share a husband than to have none at all, particularly in societies where women have less economic resources than men. So also, if there are significantly less women than men in a population, it is better for a man to share a wife than not to have a wife at all, even though he would prefer to be his wife's only husband.

In parts of the world polygamy is unacceptable from cultural, legal and religious points of view. In these societies men and women may be approximately equal in numbers, poverty and lack of resources tends not to be so extreme that significant numbers of men can't support a family, and women may not be economically dependent on men. In these societies there are no compelling reasons to have polygamy. As we saw already, few women like sharing a husband and few men like sharing a wife, so people tend not to do this if there are no economic or demographic reasons for it. **Polygamy exists because of poverty or a shortage of partners and is not the first choice of most people, it is accepted by many but only because a better option may not be available.** We can afford to have monogamous marriages if we have sufficient resources.

Another way in which marriages differ around the world is in the way that husbands and wives choose each other. In a significant number of societies prospective spouses are chosen by others. These arranged marriages, amongst other factors, take the economic status and the ethnic and religious backgrounds of the families into account. The couple who are the result of an arranged marriage may not know each other, so they may not have had the opportunity to form a relationship before their marriage. This means that neither partner in the marriage knows each other's expectations or can judge each other's dependability. However, these marriages tend to be part of extended families or live in close association with families. The expectations of the role of each spouse in the marriage and the contribution which each spouse makes to the marriage tend to be clearly understood by both families and are implicit in the marriage agreement. The families, in a sense, guarantee dependability of the spouse and also form a support network. The arranged marriage can benefit from this support network and usually also contributes to it. In fact, the agreement between the two families may give each partner in an arranged marriage a greater degree of dependability and support than two partners who choose each other and form a separate family unit. On the other hand, while an agreement between two families may be more dependable and follow strict rules, it is not as accommodating to individual differences and preferences.

People have the emotional inclination to fall in love, this is part of our emotional make-up and satisfies innate urges. There are examples from around the world of people defying social convention, even risking the penalty of death, in order to follow their passion for a certain person. Arranged marriages are generally not so much based on passion, but more on practical considerations for the couple and the two families involved. In countries, societies, and socioeconomic groups with a lack of institutional or national social support services, practical considerations are crucial to survival. Large families are the main source of support and security in a society which has little national social support. Arranged marriages can form alliances between large family groups and are therefore part of this private support system.

As in ecology in general, the balance reached when deciding on entering into marriage depends on circumstances. With marriage, there is a balance between passion and practical considerations as normally emotional and practical factors together influence the establishment of a marriage, whether it is by free choice or arranged. How important each of these factors are in the final decision depends on circumstances in the lives of the prospective marriage partners.

As we saw, while most marriages are monogamous, in many societies the option of a polygamous marriage is accepted. The Natural Religion's attitude towards polygamy is that circumstances exist where polygamy is of practical value and that polygamy consists of both polyandry and polygyny and that if one is acceptable, both should be. Of course, while polygamy is legal in many countries it is illegal in others, but it should be remembered that most people all over the world have monogamous marriages.

No one should be coerced or forced into marriage, it should be fair to the partners involved and any children in a marriage should be well cared for in every way. However, it should be acknowledged that arranged marriages play an important role in many societies and can be part of a greater family network that can give security and support. **However, part of the aims of The Natural Religion is that all people should have the resources to freely decide whom they should marry and which type of marriage they should have, uncompelled by economic or any other constraints.**

People stay together in a relationship or a marriage because they prefer it to being on their own. However, at times people decide that they do not want to continue to be part of their present marriage or relationship. There can be many reasons why they may want to end a relationship; not loving someone anymore, loving another person, or not achieving one's expectations of the relationship, can all end a marriage. Losing trust in one's partner and feeling that one is compromising too much – giving too much and receiving too little – can also be reasons why people may want to divorce and end a marriage. While many societies recognize that in certain cases marriages break down and therefore have legal divorce, it is not accepted in all societies. Some societies try to avoid the disadvantages of divorce by forbidding them. There are basic reasons why our relationships developed in our evolutionary past and these advantages, such as companionship, and emotional and practical support, are lost when the relationship in a marriage ends. In order not to lose these important facets in our lives, some feel that divorce should not be allowed. Some relationships do, however, break down. Some people do not help and support each other sufficiently, and some people do have sexual relationships with other people outside their marriage. Close emotional bonds between people can be damaged and fade away, trust can be lost, and people can act in a way that can be harmful to their partner such as physically attacking them, or infecting them with sexually transmitted diseases. Some marriage partners can also squander essential family resources, for example, if they suffer from addictions such as gambling and alcoholism.

Both partners need to cooperate and contribute to the relationship; if this is not the case, then a relationship in the strict sense of the word does not exist. Depending on how badly a relationship is damaged, both partners and children may, or may not, benefit from a breakup of the relationship. Relationships are very important and their breakup affects people's lives in fundamental ways. Situations do occur for example in which the breakup of a relationship is not in the interest of the children, although one or both of their parent may wish it. It depends on the home atmosphere and care that the children experience whether the end of their parents' relationship is better for children or not. To repair damage done to a relationship and keep an established family together can be less stressful in the long term for the parents, and better for the children, than if the parents separate. However, relationships can also break down to such an extent that it is better both for the parents and their children to end the marriage. Divorce is then the best solution in a bad situation and accepted by The Natural Religion in these circumstances.

Close relationships form a central and important part of most people's lives, so it is important that relationships are fair to the partners. The aim is that both partners have realistic expectations from the relationship, and that they help each other to realize as many of these as is possible. **So, it is advised in The Natural Religion that at the beginning of a sexual relationship, intentions for, and expectations from the relationship are honestly and openly discussed and clear to both partners.**

## 6.6: Family planning

A woman can potentially release approximately 500 eggs during the course of a normal lifetime. If all these eggs were to be fertilized, this would mean 375 years of continuous pregnancies. A man in the prime of life produces about 300 million sperm per day. It is clear from these figures, and as already briefly mentioned in Chapter 3, that no human has ever, or will ever, produce the number of children that he or she is theoretically capable of having. Evolution has shaped our bodies in such a way that biologically we can have many more children than we practically are able to have. In recent years in various countries around the world, the average number of births per mother's lifetime ranged from near 8 to less than 1.5. The indications are that in prehistoric times amongst our hunter-gatherer ancestors, a mother would typically have between 3 and 4 children during her lifetime. After we developed agriculture and during the early days of the industrial revolution families increased in size, and mothers with over 10 children were not uncommon, but even the highest of these are only a fraction of the theoretical number of children we could have. We humans do not use the vast majority of our eggs and virtually none – much less than one billionth – of our sperm for reproduction. This seems very wasteful.

So, although we are capable of reproducing during most of our adult lives, we only have children intermittently. From prehistoric times onwards people have tended to have children only when it suited them. Hunter-gatherers could not carry more than one baby, so they effectively could not have a second child until the previous one could walk well enough to keep up with the group whilst they were traveling. As people settled down in one location and began to practice agriculture, it was possible to have more than one young child as the group did not have to travel all the time. The relatively predictable food supply produced by raising plant crops and farm animals also allowed people to have more children, because they were reasonably certain that they would be able to feed them. But, even then people had far less children than their bodies could in theory produce. At different times during our history, when human population densities were high and crops were scarce, the average number of children per family fell.

In more recent times, families have also tended to decrease in size as more and more people came to live in cities. At times, in some urban societies, the average number of children per mother has fallen below 2. This means that at that time the population was not replacing itself and was therefore shrinking. In urban communities the population density tends to be high, so a shrinking population in these circumstances is not seen as a threat to society. In many richer urban societies with a low reproductive rate, food supplies tend not to be in short supply, so in these instances having fewer children is more likely to be a reaction to high population densities rather than lack of food. Comparisons of birth rates inside and outside cities have indicated that the birth rates go up as one moves further away from cities. All this means that historically we controlled and restricted the number of children we were likely to have. **Therefore family planning has been part of our human ecology for as far back in history as we can tell.**

**The conclusion drawn from this and included in The Natural Religion, is that planning when we have children and limiting the number that we have, is a necessity of life. It is simply not possible for people to give proper care to the maximum number of children which it is biologically possible for us to have.**

Humans the world over engage in sex more often for the enjoyment and intimacy of it than for fertilization, but sex can still result in pregnancy even when this is not intended and we don't have the resources to care for that child. The apparent anomaly between sex being intended more often for enjoyment than for fertilization, but nonetheless also capable of resulting in a pregnancy, is another example of an ecological balance between two extremes. One extreme is using sex exclusively for fertilization, and the other extreme is using sex exclusively for enjoyment. As with most instances of ecological balances, there are problems with both extremes.

If we have sex for fertilization only, this means that we have much less sex than we as humans are programmed to have and we could feel stress as a result of frustration due to a lack of sex. Our minds and bodies are built to have a certain amount of sex and if we don't have this, we feel an urge to have sex. Even though this is not an ideal situation, most people can deal with the frustrations due to a lack of sex, but sometimes the urge for sex can build up to such an extent that it can affect our judgment as regards when and with whom we should have sex, with possible negative consequences. Also, as sex is part of the way relationships are started and maintained, the lack of sex in a relationship may detrimentally affect that relationship. Although, as we saw before, the majority of our relationships don't involve sex, those that do are very important to us. If we only use sex for fertilization this will impact on some of the most important relationships of our lives.

The other extreme is to use sex for enjoyment only. This extreme is absolutely unstable, as we would never reproduce and the human species would go extinct. Sex developed first of all to ensure fertilization and survival of our species, but for us humans and a number of other animals species the role of sex broadened out from fertilization to include forming and maintaining relationships. Sex developed from fertilization to include close and intimate social bonding, not the other way around. So, we have to expect that frequent sex, which initially is for enjoyment only, will in most instances ultimately lead to fertilization and reproduction.

If people are having sex for enjoyment only, but a pregnancy results anyway, people may be unprepared to care for a baby. The prospect of the effort and commitment which is necessary to take care of a child for twenty years or so can cause the prospective parents to feel great stress if they are not prepared for it. In this way sex for enjoyment that unintentionally results in a pregnancy can create serious problems for both the child and the parents. The more we have sex for enjoyment only, the more unplanned pregnancies are likely to happen.

Sex which is for enjoyment only, and there being no intention of a relationship by either partner, is more likely to remain just for enjoyment if it takes place relatively infrequently. If two people engage in sex on a more frequent basis, the innate inclination to form an emotional relationship is likely to emerge in one or both partners. If the context in which sex is being used continues to be inappropriate for the formation of a relationship, this can cause emotional difficulties in the partners, the stress of unrequited love can be severe.

A new relationship may also be inappropriate for other reasons. For instance, one or both sexual partners may already be in a sexual relationship and an additional one may be completely unacceptable to their original partner and risk of the break up of the original relationship, with detrimental practical as well as emotional consequences for both partners.

As we saw before, when we engage in sex whether for enjoyment or for fertilization we may contract sexually transmitted diseases. However, when we have sex for enjoyment only, it is more likely that we will have sex with a number of different partners than if we have sex for fertilization. The more sexual partners a person has, the greater the risk of infection by sexually transmitted diseases. Once contracted, a person can pass these diseases to others with whom they have sex, either in or outside a relationship. These diseases can have very serious impact on a person's health and fertility, some even causing death.

**This book reasons that human sex has taken on a broader role in our lives than strictly for fertilization; because people want and enjoy sex when women are not fertile; because parents continue to have sex even when they do not want to have anymore children; and because of the important role that sex plays in relationships and our whole reproductive process, including rearing children. Therefore The Natural Religion concludes it is perfectly natural that people have sex for enjoyment and pleasure that they do not want to result in pregnancy.**

Because human sex has an important role in close personal relationships as well as in fertilization, we have to achieve a balance between these two functions of sex which is appropriate for the circumstances in which we find ourselves. We depend on both functions of sex during our lifetimes and we have to judge which one to use at any one time. It is also important that at any point we avoid the disadvantages of both extremes. Therefore we have to constantly assess our situation and weigh up the advantages and disadvantages of having sex and with whom. This can be difficult because when we feel sexual passion it is difficult to also coolly and objectively appraise our situation – our innate sexual feelings will increasingly influence what we do.

Yet, we subconsciously check prospective sexual partners. Those characteristics which we find attractive are often those which are symptomatic of good partners and parents. For example:

- Physical attractiveness is usually a symptom of good health, fertility, and desirable genetic traits.
- Attractive personalities are often symptoms of dependability, intelligence and general mental and emotional ability.
- Charisma, aura of power and self-assuredness are frequently symptoms of a high level of resources and social standing.

All these characteristics influence how attractive we find a person, but also whether a person is a suitable partner and parent.

Another judgment which we have to make is the timing of our types of sexual activities. When is it best to have children and when do we avoid pregnancy? If we hope to have more than one child, we also have to decide how much time there should be between them. All of these factors influence our ability to take good care of our children, and we have to try to predict whether we will be able to look after them over the next twenty or more years. Choosing our sexual partners and timing when in our lives we have children, are all essential elements of family planning and involves changing from our more usual pair bonding sex to sex for fertilization. We have to strike a balance between these two functions of sex depending on our current situation, and what we think our circumstances will be in the future.

There are a number of ways in which the human species can influence when they have children, our hunter-gatherers ancestors breast fed their children for three to four years while they travelled about gathering fruits, seeds, and roots. However, as we already saw, when walking a hunter-gatherer mother could only carry one baby or infant, so it was important that they should only have the next baby when the previous one was old enough to keep up with the group. During breast feeding a woman is less likely to conceive, as menstruation is much slower to start again after the birth of her baby than if she was not breast feeding, and when menstruation does start the first cycles tend to be infertile. Although breast feeding for three to four years did cost the mother energy, it had the effect of postponing the birth of the next child. It is thought that if a woman did became pregnant while she still had a small baby or infant and no one else could care for it, that infanticide may have been practised. Hunter-gatherer mothers may have been forced to take this very extreme and sad decision because her children's lives, as well as her own, could be endangered if she had to carry two small children. If a mother had twins, it is thought that infanticide may also have been practised. The fact that hunter-gatherer mothers did most likely resort to such an extreme course of action as infanticide, shows how crucial family planning was to their lives. Historically, infanticide was practised in many places around the world. It is thought that infanticide was used as one of the main methods of family planning in Britain as recently as the 19th century. This again indicates the pressing need for family planning, and the extremes to which people were prepared to go to achieve it.

When people started practising agriculture and living permanently in one location, average family sizes increased. The very significant rise in the world's human population at the time of the start of agriculture indicates how family planning adapted to people's circumstances. People felt they could afford to have more children, so they did.

Once human populations grew to the limits which agriculture could support, people once more started to have fewer children. People also had smaller families when bad weather conditions caused crops to fail, as happened in

Ireland during the early 1800's for instance. At different times during our agricultural history, populations waxed and waned. When the industrial revolution started, people once more felt that they could support more children and average family size increased in many regions. Cities became much bigger than before and people started to live closer together in greater densities.

Another trend in family planning is that in some societies poorer people have larger families than the richer people. This initially seems not to make sense, why do those who can afford to have more children have less, and vice versa? One explanation which is put forward for this, is that richer people are healthier than poorer people, therefore in poorer families more children are likely to die during childhood and more children are needed in order to have some of them survive to adulthood. Conditions in a richer family are more certain, so they can afford to have less children because they are more likely to survive to adulthood. A further aspect of the difference between richer and poorer family sizes is that while in richer communities families may be smaller, more people have children compared to poorer communities. So individual family sizes in poorer communities may be larger, but there are also more individuals who have no children at all, compared to richer communities. Having a large family also means that the children can help their parents make a living when the family relies on agriculture or on industrial employment for their livelihood. In recent times, particularly in areas of high population densities, the average number of children a mother has during her life has fallen. This may be a reaction to these high population densities, and also the fact that more woman are now working outside the home in paid employment.

Throughout our history the average number of children per mother has fluctuated, as the conditions in which we lived changed and as we reacted to the circumstances we found ourselves in. In the past family planning techniques included breast feeding for long periods, restricting sex during certain times, engaging in sexual activity but not in full sex, and not having sex at all. In addition, historically various devices and techniques were used to attempt to block sperm from reaching the egg. Herbs and extracts from plants have also been used to try to prevent fertilization. The use of all of these contraceptive techniques were important to the family planning of our ancestors. Apart from abstaining from sex completely, all the other techniques and strategies try to control pregnancy while still having sex. So most family planning techniques try to strike a careful balance between sex for enjoyment only and sex for fertilization only, without going to either extreme. **These are the reasons why the viewpoint regarding contraception included in The Natural Religion is, that as most of human sex is not intended for**

**fertilization, avoiding and planning pregnancies by using various contraceptive methods is a natural part of our ecology.**

Despite efforts to avoid fertilization while having sex using our modern safer and more reliable contraceptive methods, unplanned pregnancies still occur some of which are also unwanted. Estimates of the proportions of unplanned and unwanted pregnancies approximate that worldwide 50% of pregnancies are planned and wanted, 25% are unplanned but wanted and 25% are unplanned and unwanted. Depending on the circumstances of the parents, particularly the mother, an unwanted pregnancy can pose great problems. One situation in which a pregnancy is unwanted is when the pregnancy seriously threatens the health or possibly even the life of the mother. This situation does occur in a minority of cases. Other situations of unwanted pregnancies can occur as a result of rape or incest. A woman who is pregnant as a result of rape not only has the pregnancy forced on her, but also its possible health risks and the care of a child for many years afterwards. All this in addition to the physical and psychological shock and trauma of rape. A more common reason for an unwanted pregnancy is that the woman does not at that time have the resources to have a child and to take on the responsibility of its long term care.

In instances of unwanted pregnancies, historically various methods were used in different parts of the world to end the pregnancy before the child was born. Drugs, strenuous physical exercise and massages, primitive surgical procedures, magic rituals and religious prayers have all been used to attempt to induce miscarriages. The chances of success of some of these methods were uncertain and some also posed a serious risk to the health of the pregnant woman. In recent years, modern medical procedures have made ending pregnancies by aborting the ovum or foetus much more safe and certain. Presently proper medical techniques ensure that an abortion is less risk to the health of the pregnant woman than a normal birth. Safer surgical methods, in addition to changing social attitudes, have helped to greatly increase the number of abortions which are carried out each year in many regions around the world.

In the past, cultures and societies were divided over their attitude towards induced miscarriages and abortions. While some societies accepted abortions as part of family planning, others disapproved and made it a punishable offence. Currently, both within societies and between societies, attitudes continue to differ about abortions. The proportion of people who are for and against abortion varies greatly from country to country. There are also differences in opinion about when abortion is justified. Some only accept abortion to save the life of a pregnant woman, or only in cases such as rape and incest. Still others deem it to be the right of any woman who is pregnant to decide whether she wishes to continue with, or end, the pregnancy.

Most cultures and societies recognize the special nature of life, particularly human life, and those who object to abortion usually do so on the basis that it kills human life, but we have to balance our sex drive and theoretical capacity to have children with our actual capacity to have and care for children. This means we don't have a choice, but to plan how many children we hope to have and when we hope to have them. This effectively means that we limit human life before it starts. People in most communities around the world are also prepared to kill adult human beings in certain circumstances by having armies and training them to fight and kill. So, most societies are prepared to either limit or kill human life despite the fact that most people recognize the very special nature of human life. In the case of abortion, opinions differ about the stage of pregnancy at which it is acceptable. Some say that to stop implantation and to abort fertilized eggs is unacceptable, others say that a woman has the right to stop a pregnancy as long as the foetus is not able to survive outside the mother.

Most human communities strike a balance between on the one hand our acceptance of limiting the start of human life, stopping a developing foetus or killing an adult human, and on the other hand the protection of human life. The balance which most societies strike between training armies and protecting human life, is to limit the role of an army with strict rules and regulations so that killing of human life only happens in extraordinary and precisely prescribed circumstances. The balance struck between stopping life from ever developing, and creating as much of human life as we can, is to limit the number of children to a level so that we are able to take good care of all of them. Most human communities limit their numbers of children, but they can differ about the stage in the reproductive process at which this is done.

We can avoid having children by not having sex, we can also give people the option to have sex but not conceive children by means of various methods of contraception. Many societies also give people the option to stop a pregnancy either before implantation or before the foetus has developed to a stage that it can exist outside the womb, combined with the protection of an individual person's right to live from the point of birth. This balance combines limiting the quantity of human life with optimizing the quality of human life.

The system of life on earth has either got to limit itself or be limited, because all life needs resources and all resources on earth are limited. Our human ecology is part of the living system and our resources are also limited, so it is impossible for us to reproduce to our full theoretical potential, and so we can only have a limited number of children. Contraception, abortion and complete abstention from sex are all ways in which we can limit our reproduction. Most people accept that human life is precious and special and that in order to protect it and safeguard its quality, we have no choice but to limit it, but people differ about which methods they find acceptable. If we accept that sex in humans is not just for fertilization, we also need to accept contraception so that sex for enjoyment and social bonding does not result in unwanted pregnancies. In addition, a number of present day contraceptive methods do not completely guarantee that fertilization will not happen, this means that as a result of sex for enjoyment some unplanned pregnancies will happen, even when using contraception. If these unplanned pregnancies are also unwanted pregnancies, many societies give women the choice to end the pregnancy and have an abortion.

In many people's judgment, denying a woman the choice to end an unwanted pregnancy is worse than allowing her this choice. An unwanted pregnancy can have undesirable consequences for the woman's life as well as adverse influence on quality of care for that child once it is born. When a woman makes the choice to end a pregnancy, she will usually know if it is likely that she can have another child at some time in the future when she is in a better position to take care of it. This point of view concludes that if sex for enjoyment which is not intended for fertilization does result in pregnancy, it is better to have the option to carry through the intention of sex for enjoyment only, and not allow an unwanted life to develop. The root of this approach is, as we saw already, that sex in humans is not for fertilization only. Considering how much time and effort people devote to sex while a woman is not fertile, it is very unlikely that our evolution would have developed this type of sex if it did not have a purpose. It is also very clear from other animals, particularly pygmy chimpanzees, that humans are not the only species in which sex has evolved into having a broader social function, as well as being for fertilization.

Because of the importance of creating a new human being and the long term care which a child needs, the position of The Natural Religion is that it is better to have the choice to end an unwanted pregnancy with an induced miscarriage or abortion than not to have this choice. The only other option is that an unwanted baby is born, and no woman should be forced to give birth.

Of course, by far the best way to plan having children is to avoid fertilization before it happens, rather than end a pregnancy after fertilization. This book strongly encourages, and supports any measure which results in, reducing abortions to an absolute minimum, but the choice to have induced miscarriages or abortions should be freely available to any woman in the case of unwanted pregnancies, as well as in other situations such as health reasons and in cases of rape.

**The Natural Religion emphasizes the precious and amazing nature of human life, but recognizes that we also have to plan and limit human life in order to protect it. The main aim of the religious approach put forward by this book is to maximize the quality of human life on Earth.** The balance between protecting and nourishing human life on the one hand, and planning and limiting human life on the other, lies ideally in avoiding pregnancies, but includes ending pregnancies before the foetus is capable of life outside the mother. From birth onwards, The Natural Religion advocates the protection and support of all human beings, and that killing any human from this time onwards is wrong. When we have to plan and limit human life, we should do so before fertilization and in as few cases as possible also before birth, but not afterwards when we are children and adults with families and friends who love us.

A summary of The Natural Religion's attitude towards having the choice of abortion, is:

- That in certain bad situations it can be the best solution.
- That it should be freely available.
- That we should strive to reduce the number of abortions as much as possible by avoiding unwanted pregnancies, in addition of course to protecting ourselves against emergencies, such as sickness and rape, which will also reduce the need for abortions.

One aspect of women's reproductive lives, which is not seen in most other mammals, is that at a certain age women become infertile and experience the **menopause**. What makes women's menopause remarkable is, that it happens approximately between 20 and 40 years before the end of the normal course of a woman's life. Again, we need to see if the menopause has a function, because evolution tends not to result in characteristics of life which are useless. The menopause could be interpreted as detrimental, which is worse than useless, as it limits our reproductive capacity and constitutes an unproductive drain on resources. Since living species compete with each other, most biological phenomena which survive, give an advantage to the species in which it is found. Therefore, the menopause in women almost certainly gave an advantage to our species during our evolutionary history, and, since it still is a characteristic of women the world over, it most likely continues to do so.

The advantage of women's menopause to the human species would appear to be related to the care we need to give our children and grandchildren. When women have had one or more children, they usually need all their time and energy to raise these children to adulthood. To continue to have more children would seriously diminish the care they can give the children they already have, hence the need for family planning. The female menopause therefore effectively acts as an inbuilt method of family planning. It gives a woman a sufficient number of years to have children, but stops her having more at an age when she needs to concentrate on caring for the children she has, rather than having more children. The female menopause begins at around the age when a woman could also be a grandmother. Not being involved in active reproduction herself, a grandmother can, and often does, give direct help with caring for her grandchildren.

So the female menopause appears to play a role in family planning and childcare. This again indicates how important our prolonged childhood is. Sexual pair bonding, family planning, and the menopause all help towards caring for us during our long and crucial childhood.

The menopause also has a more general advantage. Human ecology is very much dependent on our knowledge information, and as we live, we gain experience and gather more knowledge. Older people have more experience and therefore tend to have more knowledge than younger people. **Therefore older women, who are not reproducing anymore, now have the time to share their experience not only with their younger close relatives but also with younger people in general. Older people therefore represent a valuable store of experience and knowledge.**

To summarize, the overall attitude towards sex included in The Natural Religion is that a balance exists between the two extremes of sex for enjoyment only, and sex for fertilization only. During a person's lifetime this balance may move towards one extreme or the other, depending on circumstances at that particular time. We can avoid the disadvantages of unwanted pregnancies, inappropriate relationships, and sexual transmitted diseases by not having any sex; but our bodies and emotions are designed to have sex, and we also obviously will not reproduce if we don't have sex. This means that while not having any sex avoids certain problems, it creates others. So realistically, the decision to have sex, or not, has to depend on the personal judgment of the potential sexual partners in their specific circumstances.

This book cannot describe all the many individual circumstances which can arise or advise on each of them, but what is pointed out here are the underlying principles of the role of sex in our lives and the basic problems associated with sex. **As general advice, The Natural Religion recommends that everyone informs themselves as much as possible about sex, so that we can assess our own situation and make informed choices about our sexual activity and that both people in a relationship should decide together when fertilization as well as enjoyment is intended, or when sex is just for enjoyment only.**

If pregnancy needs to be avoided, and effective and safe contraceptives cannot be used or are not available, then other sexual activity besides full penetrative sex can be engaged in. As long as no sperm is introduced into the woman's vagina, other sexual activity besides full sex can still give the intimacy and enjoyment of sexual contact without the risk of an unwanted pregnancy.

## 6.7: Our sexual practices

The broader role of human sex has resulted in it being very pleasurable and enjoyable. When people the world over are asked why they have sex, they are more likely to mention pleasure as the reason for having sex than the wish to have a baby. As part of human sexual activity we have developed a number of different sexual practices.

As we saw, we use sex as a fast way to form deep emotional bonds between a man and a woman who to start with are strangers. The intimacy of having sex helps to strengthen these bonds which form an important part of a loving, trusting, and dependable relationship between two sexual partners. On the other hand, people also have **sex when one or both partners do not want to have a relationship**. The sexual attraction which exist between people, the thrill of being able to attract a sexual partner, and the pleasure of sex itself can all contribute to a powerful urge to have sex with someone with whom we may not necessarily want to have a relationship. One problem with having sex for enjoyment only are sexually transmitted diseases (STDs) because people tend to have more sexual partners when they are having sex that is not part of a relationship. STDs have been part of human ecology for as long as history records. These diseases continue to have a great impact on people's lives and influence our choice of sexual practices, and so sex which is exclusively for enjoyment and outside a relationship has been prohibited and condemned in many human communities and societies and by many religions. It is of interest that in societies such as isolated island communities, where sex for enjoyment outside relationships is more widely accepted, STDs are not as prevalent compared to the rest of the world. This indicates that the underlying reasons for the condemnation of sex for enjoyment only are its attendant practical problems, such as unwanted pregnancies, inappropriate relationships and STDs, rather than that sex which is not for fertilization is inherently wrong. The position of The Natural Religion on sex outside relationships is therefore that although sex and our sexual urges are a normal part of our lives, we should take practical, emotional, and health problems into account when deciding about engaging in this type of sex.

Humans, like many other mammal species, also sexually stimulate themselves without involvement of a partner. Various methods of **masturbation** can be used by a person to sexually stimulate themselves and commonly includes a sexual orgasm. Sexual partners also engage in mutual masturbation without having full sex with the man's penis in the woman's vagina. One characteristic of individual masturbation is that it is often not spoken about and kept secret.

This is thought to be part of sexual crypsis in humans, as masturbation influences the fertility of both men and women. Sexual crypsis, as already mentioned, tends to obscure when women are at their most fertile in order to encourage regular sex between two partners. As masturbation affects fertility, part of sexual crypsis would be taken away if people knew when their partner masturbated. This therefore could be an underlying reason why people are inclined to conceal that they masturbate. Masturbation is disapproved of in many cultures, and seen as part of sex outside a relationship and therefore in some way illicit, immoral, dirty, unhealthy, a selfish pleasure, unnatural, and even a form of self-abuse. Whether this actually has the effect of reducing the number of people who masturbate is unclear, but it is another reason why people tend to conceal that they masturbate. Surveys in the western world suggest that over 90% of men and between 70% and 80% of women masturbate at some stage during their lifetime, it therefore appears that masturbation is a normal part of virtually all men and the majority of women's sexual activity. Individual masturbation only involves one person so it cannot cause problems such as unwanted pregnancies and infection by diseases. Nor can it result in unwanted or inappropriate relationships. Masturbation therefore is extremely unlikely to adversely affect another person and so it appears there is no reason why it is a mistake.

Mutual masturbation is an intimate sexual activity which can include sexual orgasms and has a very much reduced risk of unwanted pregnancies and infection by sexual transmitted diseases compared to normal sex. This means that people can benefit from the pleasure and pair-bonding aspects of mutual masturbation, without running the risk of some of the problems which are associated with sex for enjoyment. For instance, this can be a very useful manner of sexual expression for people who either don't yet have, or are at the very beginning of, a relationship.

Returning to how masturbation affects fertility, in men masturbation ejaculates sperm which are in the testes and the tubes connecting them to the penis. These sperm may be old and therefore not as active as sperm which has been produced more recently. If a man masturbates to an orgasm and then has full sex two or three days later, most of the sperm which he then ejaculates will be relatively young and active, and this makes that ejaculate more fertile than if he had not masturbated three days previously. Masturbation in women stimulates mucus production in the vagina, particularly if she masturbates to an orgasm, this extra mucus influences the survival of sperm and the ease with which they can travel into the womb if she has full sex during the following couple of days. Increased mucus production caused by masturbation also strengthens a woman's defences against invasion by disease organisms in her reproductive tract. Hence, particularly for women, masturbation can increase health, not decrease it.

Most people associate masturbation with the pleasure it brings, but that it also influences fertility and women's health is generally less widely realized. Once again this is an example of evolution developing a feeling of pleasure for something which has advantages in terms of reproduction and survival. Therefore The Natural Religion's conclusions on masturbation is that it is a normal part of most men's and women's sex lives, is completely acceptable, and has fewer problems associated with it than full sex.

Another sexual activity which many people engage in is **oral sex**. This involves one partner manipulating and sexually stimulating the other partner's genitalia by mouth. Oral sex can be part of stimulation and foreplay before full sex, as well as part of mutual masturbation which is not followed by full sex, and as such is one of a number of ways in which people sexually stimulate each other. As with masturbation, oral sex is seen by many as unnatural, unhealthy, dirty, and animal-like; however, as with masturbation many sexual partners engage in oral sex and find it very pleasurable. It is an intimate sexual activity which can be very enjoyable, and therefore has a role in pair bonding. Oral sex has an advantage in that it has no risk of unwanted pregnancy if it is not followed by full sex, but because it does involve close contact between mouth and genitalia, a number of serious diseases can be transmitted during oral sex.

It is thought that a reason why evolution has made this a pleasurable experience is that during oral sex we can closely examine the genitalia of our sexual partner, this means that we can check for any signs of disease by sight, taste, and smell. This may be very important for both partners, particularly at the beginning of a relationship, when two partners may not know each other very well yet. Oral sex during a relationship can also give indications whether a sexual partner is being unfaithful, as signs of sex and also of disease can make a person aware that their partner is having sex with someone else. While it is normally not clear exactly when a woman is at her most fertile, one way for a man to find out if a woman is menstruating and most likely infertile is during oral sex, which gets around sexual crypsis to a certain extent and may be important for him to know. So, oral sex seems to have a number of practical purposes, as well as being sexually stimulating. So the stance of The Natural Religion towards oral sex is, that it is a pleasurable sexual stimulation between two consenting adult partners, but care does need to be taken as STDs can be passed on in this way.

Sexual practices in humans are not always between members of the opposite sex, as some people also like to engage in sexual activity with members of their own sex. People who feel sexually attracted only to members of their own sex are **homosexual**, people who feel sexually attracted to both members of their own sex as well as the opposite sex are **bisexual**, and **heterosexuals** only feel attracted to members of the opposite sex. Homosexual behaviour has been seen in many animal species besides our own, including amongst insects, reptiles, birds and many species of mammals including our closest genetic relatives, the apes.

The reason why homosexuality is and always has been part of human sexual practices has up to recently baffled people. Why is a form of sexuality so persistent in us when it theoretically cannot produce offspring and so be genetically inherited? Many thought that homosexuality must therefore be learned, or a type of fashion, rather than genetically inherited. However, recent research on homosexuality has indicated that whether or not a person engages in homosexual behaviour during his or her lifetime is influenced by the genes they inherit from their parents, particularly from the mother in the case of one study. The explanation why genes for homosexuality are passed from generation to generation largely relies on a new appreciation of the role that bisexuality plays in homosexual behaviour. Homosexual behaviour is found in an approximate average 6% of men in industrialized countries but only 1 or 2% are strictly homosexual, the other 4 to 5% are bisexual. In some other cultures homosexual behaviour is found in a much higher percentage of men, because in some societies older male teenagers and young men live for a period of time in camps where all engage in homosexual activity with each other. In some cultures a prescribed rite of passage in the lives of older boys and young men was to have oral sex with the older men of the community, as it was believed that in this way young men collected the supply of sperm they would use later on in life to father children. They then had heterosexual relationships with their wives when they were deemed to be fully adult, this means that effectively 100% of men were bisexual in these communities.

The many cultures, societies, and religions around the world which accepted and encouraged homosexual and bisexual behaviour contrast sharply with many others that condemned it in the strongest of terms. Some religions are of the opinion that homosexual behaviour is sinful, unnatural, and deviant. The secular laws of societies associated with these religions often ruled homosexual behaviour as illegal and punishable by severe penalties, and it is only in relatively recent years that a number of such countries have accepted homosexual behaviour as legal. As mentioned above, it is clear from recent studies on homosexuality, that the majority of men who have homosexual feelings and inclinations, and act on them, are bisexual rather than strictly homosexual, and these men do produce offspring as a result of heterosexual sex. Bisexual activity can in this way spread genes for homosexuality in a human population.

A disadvantage of homosexual sex is that it can have a higher risk of infection from sexually transmitted diseases, as men who act homosexually tend to have sex with more partners than heterosexuals. It is also believed that anal sex, which is more often used as part of male homosexual activity than heterosexual activity, has a greater infection risk associated with it than other sexual activities. The greater risk of contracting sexually transmitted diseases is a currently accepted reason for the comparatively low level of homosexuality in industrialized countries. Societies which have much higher levels of homosexual behaviour are generally much more isolated and have fewer sexually transmitted diseases than those in the industrialized world, and therefore this disadvantage of homosexual behaviour does not apply in these populations. The industrialized countries, on the other hand, are all in contact with each other and have been for a long time, and it is in these countries that sexually transmitted diseases have been spread by all types of sexual activity.

Homosexual activity amongst women the world over is less than that of men. In industrialized countries it is approximately half, but the proportions of bisexuals and strict homosexuals amongst women appear to be similar to those of men.

In many countries with low levels of homosexuality it is still not generally socially acceptable for a woman or man to state that they have homosexual feelings. Even in societies where homosexual behaviour is legal such a statement can lead to social exclusion and discrimination, and this is a serious disadvantage of bisexuality and homosexuality. Social disapproval can be so strong that it can lead to physical, sometimes lethal, attacks on bisexuals and homosexuals. Strong social disapproval can also result in great emotional conflict in people approaching adulthood who are bisexual or homosexual. Homosexual feelings can cause great confusion, fear, and even panic in teenagers anticipating their future life in adult society. Added to this, is the confusion of bisexual desires, particularly if one does not realize that the majority of people who have homosexual feelings are actually bisexual.

One advantage of homosexual behaviour is that it can be enjoyed and can lead to strong relationships, without any risk of unwanted pregnancies. It also appears that bisexuality and homosexuality can bring a certain level of social advantage with it, because in many cultures a relatively high proportion of bisexuals and homosexuals achieve good career positions. This is usually attributed to good general intelligence and a drive to succeed, in combination with good communication and social skills.

What we know about homosexual behaviour indicates that bisexuality and homosexuality is an ordinary part of the range of human sexual preferences, and this is also the way The Natural Religion views them. Whether a small minority or a greater number of people are bisexual and homosexual can coincide, and may depend on, how common sexually transmitted diseases are in that population. Sexually transmitted diseases are a problem associated with homosexual behaviour, just as they are with heterosexual activity, but neither wanted nor unwanted pregnancies can result from homosexual activity.

In most instances when people have sex, both partners feel sexual attraction for each other, are sexually excited, and want to have sex with each other. However, if one person does not want to have sex, but the other person forces that person to have sex with them against their will then it is **rape**. Human sex is intimate and private and to be forced to have sex with someone is a gross attack on a person's freedom, self-determination, and dignity. Rape is a very serious violation of a person's privacy and the emotional effects of such a breakdown of personal safety and security can be profound and long lasting.

In most cases of rape a man forces sex on a woman. Women tend to be physically weaker than men, and are, therefore, at a disadvantage in a situation where physical violence is being threatened or used. In addition to the emotional shock of rape, a woman is often physically injured during rape and may even be killed, as well as an unwanted pregnancy possibly being forced on her as a result of rape. In other rape cases, men rape other men using anal sex. Both when women and men are raped, they can be infected with sexually transmitted diseases. Women raping men can occur but this is much less common. The Natural Religion condemns rape under any circumstances and regards it as completely wrong. Sexual activity should only happen between people if both give their informed consent. Rape is a very serious assault in a most personal and damaging way, both physically and emotionally.

A minority of people, mostly men, force sex on children. **Paedophilia** is a form of rape, and is also often accompanied by violence, threats of violence, or overwhelming emotional pressure. Normal sex is the basis of the human reproductive process and therefore is only appropriate once we are sexually mature and are able to care for children. We need to be mentally as well as physically prepared for sex and all that it entails, and children are neither physically nor mentally developed yet to have sex, they are not even able to care for themselves let alone for others. To force sex on children is therefore an extremely serious abuse of the physical, emotional, and mental superiority that an adult has over a child. Child sexual abuse is an attack on a child's normal development, with harmful consequences potentially well into adulthood. The Natural Religion's attitude therefore is that any sexual activity of an adult person with a young person is completely wrong, stating in the strongest possible terms that paedophilia is harmful, damaging, and completely inappropriate under any circumstances. Adults who feel sexual attraction towards children should never ever act on these feelings, and they should seek help from other adults who are qualified to treat paedophilia to help them deal with these feelings. The victims of child sexual abuse may also need help dealing with their memories and feelings about the abuse, and could likewise benefit from consulting those who are qualified to deal with this.

A feature of our long childhood is that for part of it our sexual organs are functional. In our teens we are not yet adult, but our sexual organs can operate. **As we mature as teenagers we increasingly experience sexual feelings and desires** and our bodies, including our minds, are preparing us for the sexual part of our adult lives. Since our bodies can have sex and we do have sexual desires, does this mean that we should have sex during our teenage years?

The purpose of sex in humans is reproduction, including the entire process from forming relationships to fertilization, followed by giving birth and rearing and caring for children. While teenagers are capable of fertilization, most do not have the resources, or are mentally prepared to care for children, in fact, they are not yet able to fully care for themselves and still need help from their parents. Most teenage girls will find pregnancies, whether wanted or unwanted, difficult to deal with, in addition to which sex tends to lead to quite deep and intimate relationships, and teenagers are not likely to be emotionally prepared for these yet. Teenagers generally rely on their parents for emotional support rather than being emotionally self-reliant and also being able to emotionally support others. The main activity of a teenager's life is learning and being educated as they do not have the knowledge and experience yet that most adults do, and this includes knowledge and experience about our sex lives. Sex is an intimate and personal activity which affects us deeply both physically and emotionally, since teenagers do not have as much experience as adults they are generally more likely to make mistakes, including regarding sex. Errors in our sex lives, such as unwanted pregnancies, relationship problems and infection by sexually transmitted diseases can all have a very serious impact on teenagers and their future adult lives.

Sexual feelings and sexual relationships develop and are part of being a teenager. As with the rest of a teenager's life, experience and knowledge about sex has to be learned, therefore it is appropriate that teenagers hear about the role of sex in our lives and also gain some experience about sex. However, a teenager's sexual experience should avoid sexual activity which has risks associated with it, such as full sex. Very intimate sexual activity such as mutual masturbation, which is not actual full sex, also has risks such as the development of very deep relationships with which a teenager is not yet fully capable of dealing, as well as disease infection. Another possible danger of mutual masturbation is that teenagers may allow their sexual passion to lead to full sex. Like the other aspects of a teenager's life, their sex life should also be allowed to develop slowly and in step with their biological and emotional development. A teenager should gain social skills on how to relate to, and communicate with, the opposite sex rather than get straight away involved in the physical part of sex. Full sexual activity, as well as other matters, is only appropriate for adults. The conclusions in The Natural Religion are therefore that, although teenagers are sexually developed, it is unlikely that they can make an informed judgment about whether sex is prudent and safe or not. Older children should be informed about the biological side of human sex as they approach the teenage years, and teenagers should be taught the emotional implications and given practice in the social skills which are associated with relations between the two sexes. People should only engage in actual physical sexual contact with each other when they are adult and able to care for children as well as themselves.

The passion of sex can be very pleasurable and enjoyable, and some people help others to feel the pleasure of sex in return for payment without any intention of this leading to any type of relationship, this is **prostitution**. Money and other resources, as well as help and assistance of various kinds, are very often given by one sexual partner to another, but this is not prostitution. The difference between prostitution and helping one's sexual partner with resources, is that prostitution does not involve personal relationships or emotional ties.

The majority of prostitutes are women, a minority are men, and they have sex with many people who are regarded as clients availing of a service and who pay money for this. The three usual risks also apply to prostitutes. Unwanted pregnancies, difficulties with relationships, and sexually transmitted diseases are all serious disadvantages of prostitution. Prostitutes are also at some risk from violent attacks by their clients, and in order to have protection against attack, prostitutes often have men close by to protect them. These men, or pimps, may also put prostitutes in contact with clients and take a part of the payment. Pimps often commercially, as well as sexually, exploit prostitutes and perpetuate the poverty and lack of choice which caused them to become prostitutes in the first place. Despite all the risks, very few societies in the world do not have prostitution, it is called the oldest profession in the world with a small minority of women deciding to be prostitutes despite the risks. Men mostly take part in prostitution as clients but some also as prostitutes. In some societies many men, sometimes a majority, have had sex with a prostitute at least once in their lives.

The main reason why women work as prostitutes is usually poverty. A number of circumstances can cause the poverty which can lead to a woman deciding to go into prostitution. The family background may be poor, she may have to care for children on her own, or she may have few, and lowly paid, employment chances. Another common reason for prostitution in urban societies is drug addiction. The effects of an addiction may make the addict unsuitable for conventional employment and drives an addict to get money by whatever means available, including prostitution. To engage in prostitution while also being an addict brings serious health risks with it. Both prostitution and addiction have their own health risks, which in combination can have an added effect, for example an addiction can greatly impair a person's capability to assess risks. The risks to the clients who take part in prostitution include sexually transmitted diseases and negative impact on their relationships. Yet, many men continue to pay for sex and take these risks so that prostitution continues to be part of human sexual behaviour in virtually all parts of the world.

Prostitution is acceptable in some cultures, but is unacceptable in many others and generally prostitutes are seen as socially inferior. Prostitution is illegal in many countries and it tends to be the prostitutes who bear most of the rigours of the law rather than their clients. If prostitution is illegal then it also tends to be associated with other crime which is to the disadvantage of everyone involved in prostitution as they tend to lose the protection which society can offer against other crime, such as being exploited economically and in terms of personal safety and health care.

The Natural Religion's assessment of prostitution is, that the most objectionable aspect of prostitution, firstly, is the poverty which drives people into this type of work rather than the fact that sex is paid for or that it involves sex for enjoyment which is not intended for a relationship, and secondly, the risks to which prostitutes are exposed in this type of activity. The aspirations on which The Natural Religion is based, include that people should have enough resources to allow them to freely decide when and with whom to have sex.

As we already saw in this chapter, in most families it is only the parents who have a sexual relationship with each other. However, exceptions do occur where close genetic relatives, such as sisters and brothers, fathers and daughters and mothers and sons have sex, which is **incest**. In the great majority of cultures and societies incest is felt to be abhorrent and is strongly condemned, although it infrequently does occur in most of these societies. In a few cultures in human history incest was accepted, such as in ruling families in ancient Egypt and in some small isolated populations in various parts of the world. In these isolated regions, people almost certainly did have sex and reproduced with close genetic relatives as they had little contact with people from outside their group and they were relatively few in number. These cases in which incest was accepted occurred in relatively small and very specific groups of people, but not in larger populations.

Incest involves having sex with close genetic relatives and children that are conceived as a result can suffer from inbreeding, which leads to ill health more often than if the parents are not closely genetically related. The reason for this is that in the majority of our populations, part of people's genetic make-up contains genes which have the potential to kill or to be seriously debilitating. Surveys of the children of genetically related parents, such as first cousins, indicate that on average a person has between one and four lethal genes in their genetic make-up. These surveys were carried out in Japan, Brazil, India, France and Britain. The cystic fibrosis gene is estimated to occur in one in 25 people in France and England. Another survey estimated that in the USA and France, on average people have four lethal genes equivalents. This means that the average American and French person has four genes which could kill him or her, or eight genes which could kill half the people that have these genes, and so on. So it does appear that dangerous genes are found in most human populations.

Why don't these genes kill all these people? The reason for this is that we have a double set of all our genes. We can carry dangerous genes without knowing it, because the other gene of the double set masks the effect of the dangerous one. In such a case the dangerous gene is called a recessive gene and the health giving gene is called the dominant gene. In fact, the reason why we have so many dangerous genes is because they are recessive, because if they were dominant that person would not have survived to reproduce and the dangerous gene would have also died out. What happens with inbreeding is that genetically closely related people have more genes in common than people who are not so closely related and this means that they are also more likely to have the same dangerous genes. So if two closely related people have a child, that child has a higher likelihood of receiving two matching dangerous recessive genes, and is less likely to inherit the dominant good genes which masks the recessive bad gene, than a child from unrelated parents. This means that inbreeding is more likely to produce an unhealthy or fatally ill child, therefore incest produces more children which are either unhealthy or die than sex between people that are not closely related.

However, if people in a population do not carry any lethal genes, then inbreeding cannot result in these combinations of dangerous genes. It is thought that in the small human populations in which a higher level of incest was accepted, dangerous recessive genes were much less prevalent than in other larger populations. Inbreeding can also occur in plants and other animals. Many flowers are constructed in such a way that the pollen of that flower is unlikely to fertilize the seeds of the same flower. In some plant species the two sexes do not occur in the same individual plant. Animal species, ranging from the great apes to horses, also avoid inbreeding by adolescent individuals of many animal species migrating away from the region where they were born or moving from their mother's group to another group. In some species it is the young males who migrate, in others the young females move away. One way in which many animals species, including us humans, counteract inbreeding is called sibling inhibition. This describes the lack of sexual attraction between adult individuals who were reared together when they were young, such as in a human family.

Ways of avoiding inbreeding and stopping incest are part of most human cultures. Incest is a punishable crime in many countries and seen by most people as a loathsome and repugnant act. Strong social pressures urge people not to marry close relatives, and in traditional cultures marriages between people from different villages not only avoid inbreeding, but also facilitate social contact for trading and political alliances.

Despite strong social disapproval of, and legal sanctions against, incest, it does take place in most regions in the world. In the case of father-daughter and the less common mother-son incest, their relationship will be different from the normal parent-child relationship. The relationship of a child with its parents is vital to the beneficial mental and emotional development of a child. If incest is part of this relationship it can influence a person well into their adulthood, and even damage their ability to form normal relationships with other children and adults later on in life. With sister-brother incest the relationship is also not what it normally should be. Generally in cases of incest in societies which condemn it, strong pressure is brought to bear by the older person on the younger person not to tell anyone about it, and this pressure can consist of threats of various kinds as well as actual physical violence. So the younger person effectively lives in a constant state of fear and intimidation. Incest between an older and more powerful, and a younger and weaker member of a family is rape – a cruel distortion of the child-parent relationship. Children need a caring parental relationship with their parents, not a sexual relationship. As already mentioned, sex at too early an age physically and emotionally injures a young person, because they are unprepared for sexual activity. Sex with children is paedophilia and is universally forbidden and the reasons why paedophilia damages children also apply to incest.

In the case of incest between consenting adults, The Natural Religion warns against this on the basis of the dangers of inbreeding to any possible offspring, and also the impact that the social unacceptability of incest will have on the people involved.

When we consider human sexual practices we see that we are amongst the more sexually active mammalian species. Sex can be extremely pleasurable for us and therefore we wish to have sex often, and for this reason **the view** included in The Natural Religion is that our wish to have sex is normal and natural and not in any way base or dirty, or something to feel embarrassed or guilty about. However, this view also stresses that sexual activity does have problems and disadvantages associated with it, and these need to be taken into account to avoid losing more than we gain by being sexually active. Some of the dangers associated with sex can adversely affect our whole future life, or even be life threatening.

## 6.8: The future of giving life

The process of creating new life has health risks for the mother and the baby. The chances of a woman dying from complications associated with pregnancy or childbirth, vary from approximately 0.02% in countries with modern health care to over 6% in countries were modern health care is not available to most women, and the chances of a baby or infant dying range from less than 1% to over 10% respectively. These figures show that modern health care and medical techniques have made giving life and having babies safer than it used to be. This trend will hopefully continue in the future. In addition to good health care, modern contraceptive methods have also become safer and more reliable to use, and this means that parents are better able to plan when to have their children. Parents who can choose when to have children tend to do so when they are best able to care for them, and these children are therefore more likely to receive better care than if they had not been planned. If a child receives better care then it is likely to be healthier and survive better, and lessens the need for parents to have more children. In the future our reproduction should therefore be less wasteful of human life, and safer and more predictable for mothers and their babies. Therefore The Natural Religion sees these new techniques as positive and beneficial and supports the further advances that are needed. However, one concern is that these advances in medicine are not as available to poorer regions as they are in richer regions. **It is more dangerous both to have a baby as well as be born, in a poorer less industrially developed country than in a richer country. To make advances in medicine available to all, irrespective of what country we live in, is one important aim of The Natural Religion.** All aspects of family planning should be carried out using safe medical procedures and health care methods, and all women should be in a position to make free and informed choices.

It is now possible to carry out tests which show whether a foetus is male or female. Some people are alarmed by the possibility of parents terminating pregnancy in countries where abortions are allowed if the foetus is not the sex which the parents wish for. Opinions differ whether it is acceptable to terminate pregnancies for these reasons and fears are expressed that this could result in many more children being born of one sex than another.

Evolution has shaped us to have approximately equal numbers of baby girls as boys. The reason for this is that if there are more of one sex than the other, then the less numerous sex is more in demand reproductively. This, so to speak, follows the supply and demand rule in the reproductive market place. If certain genes are more in demand, they will be more successful reproductively and therefore their genes will be inherited more by future generations. This has the effect of the less abundant sex taking over from the previously abundant one, but then the advantage swings in the other direction and the same thing happens in reverse. Over many generations the numbers of the two sexes reach a balance and this is thought to be the reason why the sexes are approximately equal in numbers in most animal species.

If we apply this same principle to the situation where, as a result of parents choosing the sex of their baby, one sex was more numerous than the other, the least common sex would become more popular because it would be in short supply. It is therefore probable that effectively the same balancing process as in evolution would happen if parents were to choose the sex of their offspring. So, whether we have the knowledge and technology to choose the sex of our babies, it probably would not make that much difference to our human sex ratio in the medium or long term.

Medical tests can also indicate whether the foetus has certain disabilities or whether its health is threatened in certain ways and this allows for the possibility of remedial action, either during pregnancy or shortly after the birth. If the foetus is disabled and no remedy is possible, parents in countries where abortions are accepted have the option to end a pregnancy. This course of action is criticized by some.

The majority of people value and protect human life. As we know, for some the balance between family planning and creating and protecting life does not include abortion. However, many other people are of the opinion that abortion should be part of family planning, and do think that it is better to have the option to abort a foetus which is seriously disabled and would have a fundamentally impaired life. In our evolutionary past only slightly disabled people could survive with the help of others, for example, presently most hunter-gatherers are generally quite healthy because their life style does not allow disabled people to survive childhood. However with modern medical advances, more disabled people can be helped to survive and given an improved quality of life. This help for substantially disabled people needs to be available for the entire lifetime of a person in this situation. Society at large should give this type of long-term care as parents grow old and other members of the family may not be in a position, or have the resources, to provide this.

The Natural Religion's position on all matters concerning family planning is that in the interest of good childcare and giving a child a good quality of life, all options should be available to parents including the ending of a pregnancy. **Therefore, if parents wish to abort a foetus which is seriously disabled this should be an available choice, just as much as having the choice, made possible by long-term support from society in general, to have and care for a disabled child.** This should be taken into account both when planning a family and also when a society is planning for care of its disabled members. The Natural Religion's answer to those who deplore the denial of the chance of life of a disabled foetus is, that it gives parents the option to plan and give the chance of life to another able-bodied child.

A number of new medical techniques are being developed which have the potential of influencing the future of giving life. Disability causing, or lethal, genes occur in most human populations and as more is discovered about these genes, it is becoming possible to detect if a foetus or young baby inherited them. The potential exists to either suppress or mask the functioning of these genes, or replace them by means of gene therapy.

Gene therapy also has the potential of counteracting cancerous growths in people of all ages. The reason why cancer cells grow out of control lies in the genes of these cells, and gene therapy can potentially correct the malfunctioning of these genes. As the genes in the nucleus of our cells control the way they function, our whole body completely depends on our genes functioning correctly. Considering how crucial our genes are to us, worries have been expressed about possible mistakes made with gene therapy.

Another medical technique which has aroused considerable mistrust is human cloning. This involves taking a cell from a person and treating it in such a way that it grows to form another person, and these two people would then have the same genes. Genetic intervention and cloning are not new in that they effectively occur already naturally. Viruses infect us by injecting genetic material into our cells which alters the functioning of these cells. The effect of the viral genetic material include causing the infected cells to produce new viruses. Infection by a disease-causing virus may not be therapeutic, but it is genetic intervention. So also, nature produces clones every time identical twins are born. In the world's human population, which is now more than 6600 million people, approximately 45 million people are identical twins. Identical twins arise when a single fertilized ovum splits and these two cells separate to form two individuals, rather than staying together to form one foetus. If no further mutations occur early in the development of either ovum, the twins will have identical genes and therefore they are clones. The mistrust, and at times fear, of both gene therapy and human cloning is probably due to the essential role of our genes to our existence. If we make a mistake with the way we use gene therapy it could theoretically have serious consequences. However, these methods have a great potential for helping people with health problems, improving people's quality of life, and saving lives.

In the case of gene therapy, controls need to be enforced to avoid that this method is used to harm people, similar to the ways biological warfare is used to harm people. Research and development in this field needs to be open and accessible to scrutiny, so that its progress and any possible misuses can be noticed in time. One of the great advantages of gene therapy to counteract diseases is that it may in certain cases reduce our dependency on drugs such as antibiotics, as well as cure currently incurable diseases.

Cloning used in a normal family setting would not produce two persons which are exactly the same. In fact, even as two identical twins grow in the womb, they experience differences in circumstances because the supply of nutrients from the mother's placenta may not be exactly the same to each twin. As we saw before, every living being is the result of the interaction between their genes and their environment. A good way to explain this is to liken us to a cake, which is made using certain ingredients as well as an oven in which to bake it. The ingredients are like our genes and the hot oven is like the environment in which we grow up. Both the ingredients as well as the oven are needed to make a cake, neither the ingredients in their own, nor the oven on its own is a cake. So also with every person, our genes need to interact with our environment to form us. Therefore, two people with identical genes will not be exactly the same, because the circumstances in which they grew up are never exactly the same. Not even identical twins who developed in the same womb, and who are reared in the same family are exactly the same, because they cannot occupy the same physical space at the same time.

**Support for gene therapy is part of The Natural Religion, because of it's great potential for helping people who are suffering from serious diseases.** Previous warnings are repeated here, that as with any other powerful technologies, genetic interventions of various types should be easily scrutinized and monitored. **The conclusions regarding human cloning included in The Natural Religion are that since identical twins are genetic clones, and that as a group they do not represent any threat, that human cloning per see is not a threat.**

The potentially new medical technique of xenotransplantation is not directly concerned with human reproduction, but is mentioned here as it is often discussed in conjunction with gene therapy. These transplants of organs from other animals into humans have the potential to save many lives and improve the quality of life for many more people. The main problem with transplanting an organ even from one human to another is that the body of the person receiving the organ may reject it, because it perceives it as a disease-causing foreign body. However, if the genes of another animal could be changed in such a way that a person's body would not reject an organ from that animal, we would have a greatly increased supply of organs available to us to cure people. So, one way to make other animal species' organs compatible with ourselves would be to change those animals genetically.

Another possible way to produce tissues and organs that a person's body would not reject, can come about as a result of human stem cell research. This area of research investigates adult as well as early embryonic and umbilical cord cells, and is permitted in some countries but illegal in others. The potential in stem cell research is that these cells can change into cells from any organ. This means that if any organ is misfunctioning or damaged, and if stem cells are taken from the same person's body, these cells could theoretically repair the organ without giving rise to rejection problems.

Changing genes of other animal species has been carried out by people for thousands of years by selectively breeding with those individuals who most closely approach our requirements. However, people have more reservations about changing genes using laboratory based genetic research techniques. Again, the fear of changing something as basic to life as genes also influences peoples' view of xenotransplantation and stem cell research, so this type of development would also need to be monitored and be open to all for scrutiny.

The alternative is to forbid this kind of work. Such a ban, if effective, could cut humankind off from the potential benefits of genetic alterations. In addition to this, despite legal bans on developing techniques which alter genes, this work could still be carried out illegally or in other countries. There are great incentives to develop and carry out techniques which take away, add, and change genes, whether they are legal or not. Work on changing genes carried out in secret can not be easily controlled or monitored. Possible dangers due to genetic alteration could therefore develop, either by mistake or on purpose, which would then be hidden and therefore maybe not be noticed and corrected in time. So a legal ban on this type of research could have the opposite effect to what was intended. It will be more likely that laboratory based work on alteration of genes will proceed in a beneficial way, if it is carried out legally and open to all for monitoring and scrutiny, than if it is carried out illegally.

The potential, of the new technique of xenotransplantation of organs for improving the health and quality of life for many people, is great. As long as the animals which would be used for this were given full and proper care and did not suffer, making their organs compatible with human bodies, so they could help many seriously ill people, certainly seems a good development. Human stem cell research also has great potential to lead to the relief of human suffering. **For these reasons, The Natural Religion supports development of xenotransplantation and stem cell research, and sees no sense in banning them, and encourages open scrutiny and monitoring of these techniques.**

As we discussed in this chapter, our relationships and sexual life is fundamental to our reproduction and continued existence of us as a species. For this reason, religions have traditionally advised on this part of our lives. The approach incorporated in The Natural Religion is to see the way we humans give life as basically similar to other mammal species. **The difference between us and other species is primarily due to the fact that we have a more developed brain combined with the fact that we walk upright, which means that baby and childcare is intensive, of long duration, and crucial to our full development.** A close relationship between the parents promotes good childcare; and human sex, besides leading to fertilization, is also for maintaining this relationship. We have to strike a balance between having sex for fertilization, and sex just for enjoyment, and avoid fertilization when this is not intended. **The Natural Religion views the entire process, including not actually having children but helping to create a suitable environment for other people's children, as the way in which we humans reproduce and give life.**

People's individual efforts in reproducing our species have been extremely successful. This has led to an unprecedented increase in Earth's human population, particularly in recent times. A sharp rise in population always influences the ecology of any type of living organism, including ourselves. So in the next chapter, we will have a closer look at our human population boom and its implications.

# Chapter 7

# Populating the Planet

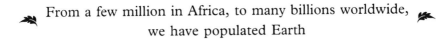

From a few million in Africa, to many billions worldwide,
we have populated Earth

## 7.1: Success of the human species

When our ancestors were hunter-gatherers it is estimated that there were less than 5 million people worldwide, only reaching this population size 10,000 years ago. 7000 years later, or 3000 years ago, it is estimated that the world's human population had increased ten times to 50 million. It was during these 7000 years that we developed agriculture.

By 2000 years ago, 1000 years later, we reached 130–250 million.
By 1600 A.D., 1600 years later, we reached 500 million.
By 1830, 230 years later, we reached 1 billion (1000 million).
By 1930, 100 years later, we reached 2 billion.
By 1960, 30 years later, we reached 3 billion.
By the mid 1970s, 15 years later, we reached 4 billion.
In 1986/87, 17 years later, we reached 5 billion.
By the year 2000, 13 years later, we reached 6 billion.

So, from the time that most people changed from living as hunter-gatherers to relying on agriculture for their food, our world population rose from 5 million to 6000 million – a 20,000% increase! It took 90,000 years for our population to rise to 5 million living as hunter-gatherers, while it only took 10,000 years for us to rise to 6000 million (6 billion) once most people came to rely on agriculture.

During the time that agriculture became part of human ecology, the rate at which our population increased has risen enormously. It took 7000 years for the number of people on Earth to multiply by 10. Then the population multiplied by 10 in just 2600 years. After that and up to the present, our numbers increased more than 6 times in just 400 years. This is not just increasing our population, but increasing the rate of increase in our worldwide population.

The sheer size of our population – the total number of human beings – is so vast that it is virtually impossible to visualize or comprehend. Furthermore, it is not only the total population which is so big, but the numbers of people which are being added to our population are also huge and increasing. During the years between 1960 and 1995 the world's human population was increasing by an average of over 70 million people per year! That is the whole human population of the world of about 2500 years ago being added every year. During the five years from 1995 to 2000, the population increased by 100 million people each year!

From the mid 1970s to the year 2000 the world population increased by 2 billion people, that is as much as the entire human population of the world around 1930. During the last 10 years of the twentieth century the equivalent of Earth's entire human population of 300 years ago was added to our population. During the last 5 years of the twentieth century the entire human population of 400 years ago was added to our population. By 2003 we reached 6.3 billion; and one estimate for the world's human population in November 2007 was more than 6.6 billion, so the trend of 100 million people being added to our population each year is continuing.

Never ever in the history of the human species has our population increased so much in one generation as it did during the last 50 years. If a population increases by just 2.8% per year, then that population will double in 27 years. When such increases happen in a population which is already big, then the addition of people to that population will also be very large.

Predictions vary about future population increases. Some estimate that our human world population will stop growing at somewhere between 7.9 and 11.9 billion people during the twenty first century. These forecasts take into account that in 2000 A.D. in some regions in the world between 37% and 45% of people were less than 15 years of age. These young people are likely to have children during the subsequent 30 years, so we have to expect further increases in our population.

If we keep reproducing at the rate we were in 1990, we will reach a figure of 700 billion by the mid 22nd century. This means that in 150 years' time we would need 116 times as much food, water, shelter, energy and other resources as we now use. In fact, during the course of the last 3000 years we have already managed to multiply our resources by approximately the same figure to meet the needs of our rising population. So we would theoretically have to change our future human ecology at least as much in the next 150 years as we changed it during the last 3000 years in order to support such a population.

Many forecasters, however, think that we may not have enough food, water and other resources to sustain our present population level, much less a 116 times increase in human population. Some predict that we will reach a population peak in about 2030 A.D., after which our population will fall by as much as five-sixths.

**So, agriculture is the reason why so many people can be alive today. Our knowledge has allowed us to live in climates and locations for which we were not originally suited and we have spread out from Africa – our evolutionary homeland – over virtually the entire land area of our planet.** Using knowledge, we developed the technology to make clothes to wear and shelters in which to live. Clothes and shelters protect us from climates which are otherwise too cold for us.

It is therefore concluded in The Natural Religion that our capacity to use knowledge has given rise to our huge population increase and demonstrates the amazing evolutionary success which knowledge information has been so far.

## 7.2: Dangers of our success

The success of a species of living organisms is measured by how it can increase, spread, and maintain its population. As we saw in Chapter 2, life is essentially about gathering the sun's energy, locking it into chemicals, and using it to maintain oneself and help reproduce one's species. While maintaining our population is biologically a success, increasing and spreading our population is being more successful. Most living species have the capacity to increase their populations when the opportunity presents itself. If the supply of food and resources stays more or less constant, then the population numbers should also stay much the same, but if new resources open up it is in the evolutionary interest of a species to increase their number so they can use those resources before another species uses them. Our human reproduction can do this just like most other living species.

As we have already seen, **the best way to summarize how the ecology of a species works is, that a balance exists between two extremes, which**

**shifts somewhat towards one extreme or the other in order to deal with changes in circumstances.** The number of individuals in a population of any species is one of the most important factors influencing its success and affecting its ecology in general, and here also balances exist between two population extremes. If there are too few individuals in a population reproduction may be very inefficient or not possible. Also, if a population is too small, it may lose out in the competition for resources to other species by being outnumbered. On the other hand, if the population is too big other problems may arise, the first of which is usually a shortage of food threatening wide scale death due to famine.

The most stable and desirable population situation is, therefore, a balance between a low and a high population. If resources are scarce then the balance has to move towards the low population extreme, and if resources are plentiful then a population can increase. On a worldwide basis, we humans have been extremely successful in increasing our population, but there have been many instances in human history of local overpopulation resulting in suffering and death.

There is a typical sequence of events leading up to an overpopulation of any living species, plant or animal. In the initial stages of population growth there are few individuals, and food and other resources are widely available. At this stage, the growth of the population is limited more by how many offspring it is biologically possible to have per generation, rather than by the amount of available food. As long as there is enough food, the population can usually increase. One aspect of growth in any population is that it increases in proportion to the size of the population which is already there. Compare two human populations, one big and one small, both increasing by 1% per year. The number of people added to the big population is much more than the number of people added to the small population; an increase of 1% to a big population needs much more additional food than 1% increase to a small population. Therefore in the initial stages of a small but growing population, the increases per generation tend to result in a relatively small additional demand on the food supply. As the population grows the increases per generation start to get bigger. At some stage, and as each succeeding generation adds greater numbers of individuals to the population, the food supply and other resources start to get used up. When the population is big, food supplies may already be stretched and just one additional generation can have the effect of causing a serious food shortage. Typically as competition for food gets greater, individuals may not have as many offspring as when there was little or no competition for food. In the unlikely event that reproduction were to stop completely there still remains a large population which have to be fed in the short and medium term. So, as a population grows and reaches the limits of its food supplies, its reproduction rate usually lessens. At this stage it depends on how reliable the food supply is as to what happens next.

If the food supply stays stable the population may stay around a level just short of using up all its food, and many plant and animal species have developed such a level of reproduction. There are many examples of animals such as rabbits, deer, rats, and tree shrews, which react to high population densities by reducing their reproduction before food becomes scarce. This reaction seems to depend on how many animals live close together rather than how much food is available at that time. We humans can react in similar ways. As already mentioned, surveys have shown that in the case of some cities that the average number of children per mother rises as one moves further away from the city centre into the surrounding countryside. In these cities and their surroundings there are no food shortages, it would seem therefore that we also react to large numbers of people around us with having fewer children even if we have enough food.

Food supplies vary for different reasons. For example, a herbivorous animal species, such as deer, living in a temperate climate which reproduces once a year during spring and early summer, will have a seasonal food supply with enough food during the spring, summer, and autumn to feed the whole population and the new offspring of that year. However, if the population is so large that they all eat their food by midwinter, then the whole population may suffer from lack of food by late winter and early spring. Usually in this case, the weaker members of a population will die, and those that are stronger are able to survive until food once more becomes available later in spring.

Food supplies often depend on the weather. When weather patterns change for one or a number of years, food supplies can decrease below that which a population needs over the course of a full year. Food supplies can also be reduced just by a population using them up. If grazing animals are so numerous that they eat and kill the plants which they depend on for food, they will have no food in the future. In human history there are many examples of there being so little food that the seed for the following season's crops was also eaten, making the famine even worse. Degraded soils due to over-use and bad, non-sustainable, farm practices is another reason for a lack of food. Large populations need a lot of food, and an unrelenting demand for food from growing populations can lead to unsustainable farming.

Certain events commonly take place when a large population runs short of food. First the individuals of the population weakens due to lack of food. In our own case, when people are underfed and weak, our resistance to diseases drops and many become sick. When large numbers of people are living close together they can easily infect each other. In this way tens of thousands and sometimes as many as millions of people die from diseases in dense populations which are weakened by lack of food. In China and India during the 19th century, tens of millions of people died from lack of food. It is estimated that in recent years between 10 and 30 million people die world-wide each year as a result of lack of food and over 500 million people are undernourished. When food and other resources run short, competition and aggression often also result. This happens amongst humans just as it does amongst other animals, and wars and conflicts are often part of, and con-tribute to, the impact of famines by destroying crops and infrastructure. This series of events has been repeated many times over during the course of human history.

Our worldwide human population expansion can be seen as a great success for us as a species. Other animals, like ants, bees, and termites also create their own living conditions, and live in high densities. They have genetically developed this way of life over many millions of years. We humans have found ways to do the same, but in a fraction of that time by using knowledge. Taking an overall view of our species, by creating more people we also seem to have increased our capacity to support more people. Our higher populations seem to have brought about an ever increasing inventiveness to cater for bigger populations. The rises in population in the last few millennia have gone hand in hand with advances in technologies, ranging from agriculture to health care, and from mechanization to computerization. This could give us the impression that as long as we keep advancing our knowledge and technology, we can also keep increasing our population.

We humans are living organisms. All living beings, whether plants, animals, bacteria, or viruses need resources and no resource on Earth is limitless. Therefore, no matter how many new resources we find for ourselves, and no matter how much we can increase our efficiency of using already existing resources, a limit will always be reached. While an increasing and spreading population is the sign of a successful species, no species can keep increasing in numbers for ever. Too much success can be very dangerous for any living species. Overpopulation can kill millions and has done so many times in the past.

Our current world population has never ever been bigger, this means that recent increases in our world population have also never been bigger. As we saw, during the last 30 years of the 20th century the number of people which were added to our population was the same as the entire world's human population in 1930. Our species has never ever experienced such increases before: we are living through the first time this has happened with us humans. In the short term we have to expect even further increases in numbers of people, because many of our populations consist of large proportions of young people who will have children in the near future. All these people will need food and other resources to avoid possible disaster, and because our population is bigger than ever, future disasters could also be even bigger than the catastrophe of the deaths of tens of millions of people as happened before.

**Currently we have to accept that the potential exists for a human calamity on a huge scale. The reason for this is the rate of our population increase during the latter half of the twentieth century.** If the ecology of any living species shows as fast an increase in population as ours has during the last 100 years, the normal outcome is a decline in population. The steeper the rise in population, the more likely it is that a sudden catastrophic population crash rather than a gradual decline will be the outcome.

As worldwide there are now more and more people, our increases in population are getting greater in shorter periods of time, and this means that if we run out of food this is likely to happen increasingly suddenly as our population grows. **If we don't control the number of people on Earth to suit our resources ourselves, then famine, war, and disease will do this for us.**

For these reasons The Natural Religion adds its voice to those who warn that our success as a species can also be a great danger to us if we let our populations get too big. We have succeeded in greatly increasing our food supply and other necessary resources. In addition, we have made great advances in technology during the 20th century, particularly in the area of medicine. A combination of such reasons have allowed us to reach the unprecedented high population that we now have, but this level of increase cannot continue for ever. Even if we were to be able to create new resources in the future on a similar scale as the way we created new and increased resources in the last 10,000 years, there will always be a limit that we cannot go beyond.

The Natural Religion's position on the size of our population is therefore that there is always an upper limit beyond which we should not rise. The available amount of food and other resources determine how high that limit is at any one time and the target for our population should be below that limit. We should control our human population in benign ways that do not cause misery and death; because if we don't, we have to expect that it will be controlled in ways which will cause much suffering and tragic waste of human life. **Part of The Natural Religion is the conclusion, that we humans do possess the knowledge to enable us to match our population numbers with our resources in ways that are compassionate and avoid human suffering.**

## 7.3: **Matching population size to our resources**

As we know, our store of knowledge led to agriculture and medicine, and further technological advances in both these and other areas. All these new methods and techniques vastly increased food production, gave us other resources, allowed the large increases in our population, and continues to support our world population of over 6.6 billion people. In history, towns and cities and other centres of population depended on agricultural production to feed its people, in fact centres of civilization could only develop in those locations where there was sufficient food available for its people. So also, many civilizations collapsed because their food supply became scarce. In human culture a typical sequence of events is as follows. Firstly, rich arable land attracts people to grow food crops and rear farm animals. Initially, these people produce more food than they need, so they can have many children and more people join them from elsewhere. The population grows and as the land is not yet fully used, more people grow food and the food surpluses also grow. At this stage, towns and cities develop which are populated by people who don't grow food themselves, but rely on others to grow it for them. As long as the farmers can produce a surplus of food the cities can exist and continue to grow, and now many thousands of people rely on food produced by others.

What has then happened in many instances where land was being used to feed urban populations, is that intensive growth of crops and removal of these crops from the land took away the soil's nutrients. This meant that agricultural production fell. Soils also degrade in other ways, such as when they become too salty for crop growth due to evaporation of irrigation water, or when erosion by water and wind carries it away. The larger the population depending on an area of land, the greater the pressure on that land and the greater the chances that the soil will degrade. In these situations famine can be at its worst. **A saying which is sometimes quoted, is that forests precede civilizations and deserts follow them.**

Large populations also produce a lot of waste. These waste products can harm people directly or indirectly by affecting the water and soil we use for agriculture. Particularly since industrialization, people have produced waste products which are very poisonous. Also since industrialization, many more people exist on Earth giving rise to more waste. Initially we did not fully appreciate the effect of some of these dangerous wastes, but as industrialization developed we learned more about their harmful effects. Some of our wastes are even suspected of influencing the normal changes in climate which occur on this planet. Any climatic changes can have far-reaching effects on the growth of our food crops, and global warming will also cause sea levels to rise, threatening those people who live in and depend on low-lying coastal regions. Some of our waste gases are thought to contribute to global warming.

All possible changes in climate caused by human activity need to be put in context with the periodic cycles in climate which occur on Earth anyway. At the moment, and approximately for the last 10,000 years, Earth's climate has been warmer than during the last ice-age. If cycles in climate continue the way they have been over the last few million years, we are now between two ice-ages, or in an interglacial period. So even though polar glaciers are currently shrinking, theoretically we have to expect another ice-age soon, as interglacial periods have tended to last between 10,000 and 15,000 years. In addition to this, there have been periods of years during the 14th century and again in the late 17th and 18th centuries, when our current interglacial climates worsened and crops failed. These so-called mini ice-ages are thought to have been the root cause of famine, violent conflict, major outbreaks of disease, and drastic reductions in populations in many parts of the world.

**While there is evidence that human activity has influenced climate change, we also have to expect the climate on Earth to worsen due to non-human factors. When this happens we will not be able to grow as much food as we can at the present using current technologies; so, if we want to have as many people living on this planet as we have now, then we will have to find new ways of producing food.** Another option is to manage our human population numbers to suit what we think our resources will be. This means predicting our future food production and controlling our population to a level near but below the maximum number of people which our predicted food supply could support.

The Natural Religion's main aim is to cherish human life and optimize quality of life for all, therefore, as much information as possible should be gathered about the abundance of our future food resources and climatic change. It is up to us to make sure that our future generations will have enough food and other resources to survive, and what our current knowledge indicates is that as well as developing new techniques for sustainable food production, we will also have to plan our population size to be somewhat below the level of what we estimate our future resources will support.

## 7.4: Are we suited to live in high densities?

All mammals can be called social animals because males and females have to come together to reproduce. The behaviour which is needed for the two sexes to come together and have sex successfully so an egg is fertilized is, in the strict sense of the word, social behaviour. Even if an animal lives on its own in a territory, it still has to communicate with the individuals in the territory next to it to signal where the boundaries lie, this is also social behaviour. However, in a number of mammal species social behaviour includes much more than what is needed for successful reproduction or establishing a territory. We humans were not herd animals, as during our evolutionary history we lived in groups of approximately 15 to 40 individuals and required social skills suitable for groups of this size. Therefore, most of us will have received a genetic inheritance which has adapted us, emotionally and otherwise, to living in small to medium sized groups of people.

As we saw in Chapter 4, with the development of agriculture we stopped moving around because we had enough food to settle down and stay in one location. The efficiency with which agriculture produced food allowed settlements of people to grow to many times the size of hunter-gatherers groups, giving rise to urbanization and villages, towns, and cities. Besides agriculture, another factor became important in the way people started to live together in large numbers. The people who did not have to produce food themselves had the time and opportunity to specialize in other work. At first this may have been work that they traditionally had carried out as part of farming, or hunting and gathering, but now people could concentrate on this work all the time which had the effect of making them more efficient and so producing more products. A system developed whereby farmers gave the towns' people food and received their products and services in return. **The principle that people specialize in, and become more efficient at, certain types of work, has been the backbone of the success of towns and cities and resulted in them springing up in most populated areas of the world.**

Those of us who live in urban centres, only know a small fraction of the people in that town or city personally. In terms of the population density around us this is more like the way herd animals live rather than the way we used to live, but the number of people we actually know personally is still more like the group size of our evolutionary ancestors. Previous to the 19th century it is estimated that approximately only 2% of the world's human population lived in an urban setting; nowadays approximately half our world population are urban. So, very many people are living in high densities, while we did not live as part of such large groups when evolution was shaping us. This is sometimes thought to be part of the reason why in earlier times cities were often quite unhealthy places to live. Death from disease was commonplace in cities and emigration from rural areas to cities was important in maintaining urban populations. Now urban areas maintain and increase their populations without relying on immigration. Cleaner living conditions, better hygiene, and modern medicine have greatly reduced the dangers from disease. On the other hand, in certain areas of cities antisocial behaviour can be a serious problem, violent crime and strife between people are often attributed to too many of us living too close together. However, there are areas of high population density which do not have more problems and conflict per head of population than regions of low population density, which would indicate that it is not high population density alone which causes these problems. The organization of any society, and the manner in which it is managed, is critical in determining what a society is like to live in, and people have proved that we can live together peacefully in large numbers even though our evolutionary ancestors lived in much smaller groups. We can make cities peaceful and healthy places in which to live by using knowledge to shape and manage our urban societies.

Unlike other animals, our urban life is based on knowledge information not on our genetic information. Social insects like bees, ants, wasps, and termites are genetically programmed to live in large colonies in which they carry out specific roles. These colonies and nests are effectively the same as our cities with certain individuals carrying out specific tasks while others are specialized in other activities. Species of ants of the genus *Atta*, found from the southern USA to Argentina, have gardens in which they grow food to feed the other ants of their colony. Similarly, people in our cities depend on another group of people to produce food for them. The social insects are a very successful group of animals and because of their efficiency often represent a major proportion of the living animal biomass in their habitat. Humans have also discovered the efficiency which specialization of work brings. The specialization of food production in human agriculture has allowed other people to specialize in other roles, raising the efficiency of the population as a whole. So, we have therefore increased our efficiency using the same basic method as the social insects, the difference being that social insects have evolved specialization by means of direct genetic control, while human specialization is directly dependent on our knowledge information. Being directly based on knowledge, we are much more flexible in the way we live. Unlike social insects we can live in small extended family groups, but we can also live as part of a city together with hundreds of thousands of other people. We can carry out many different tasks during our working lives, or we can just concentrate on one specialized job. We can work as part of a large organized team of people, or we can function as a one-person operation. Behaviour under direct genetic control such as in social insects, does not have this flexibility.

Being based on knowledge, we humans, therefore, also need to use knowledge to manage our urban societies and deal with the problems that thousands of people living closely together encounter. Because, while living closely together has advantages, it also has disadvantages. We should remember that while many other animals have always lived in large groups, this is a very recent strategy in human ecology. **Currently urbanization is a worldwide**

**success because our cities are growing not shrinking, however urbanization is a recent trend and its current success does not mean that it will continue to be a success.**

There are two basic problems associated with urbanization.

The first problem is that of disease, conflict, and strife. Any animal species, particularly those which are not genetically adapted to live in large groups, will run the risk of disease epidemics when living in high densities. Most animals will also become more aggressive towards each other when competing for resources with many others. In human history, disease epidemics have decimated urban populations on many occasions, and in the past, and continuing to the present day, violence and crime are problems in many cities around the world. However, we have made inroads against the twin problems of disease and aggression. Hygiene, modern medicine, and generally improved health care have made living in cities much safer than it used to be. Aggression and crime are not big problems in all cities, therefore, there are ways in which societies can be organised so that violent conflict is reduced to a minimum.

The second problem is more fundamental than the first one. This concerns the specialization which has made urban life so powerful. Specialization in the ecology of any living species usually brings greater efficiency and success with it. However, as we discussed in Chapter 2, specialization also tends to brings loss of adaptability. Generally speaking a very specialized species cannot adapt to a change in circumstances as well as a less specialized species. If the environment changes too fast for a species to adapt, that species' survival is threatened in a very fundamental way.

Specialization in our urban life has made those of us who live in cities very efficient, and this is one of the reasons why we can have so many people on Earth. Specialization means that everyone in a community is dependent on other members of that community to give them essential resources, in other words, there is a high level of mutual dependence and a low level of self-sufficiency. **A high level of specialization and mutual dependence is very efficient and can support millions of people when it works, but if this mutual dependence for some reason does not work anymore, all these people living closely together still need essential resources and at the same time. This can result in a disaster of truly appalling proportions.** As we saw in Chapter 2, while specialization leads to greater efficiency and bigger populations, it also leads to loss of adaptability and increased risk of failure when conditions change.

At all stages of the development of urban life in human ecology, the urban centres of population depend on rural areas to produce food for them. Soils can and do erode and/or degrade, climates do change, water supplies can run out, and political situations change and can cut off food supplies. Any of these problems can cause food supplies to dwindle or stop completely and the urban centres depending on this food will then also diminish or disappear, as we know from our history. We cannot have our specialization and our mutual dependent activities, if we don't have food to fuel this system.

As mentioned earlier, energy has become increasingly important in the life of our cities and urban centres. Transport (particularly of food), heating, lighting, all of our utilities, and most of our services depend on energy. All of these make safe urban living possible. Much of the energy which we use on a worldwide basis comes from the fossil fuels – oil, coal, and gas – or is converted into electricity from one of them. These three sources all have a limit to their supply and at some stage they will run out. Hydroelectric power will be there as long as the water turning the turbines keeps flowing, but the amount of electricity which is generated using hydroelectric dams is only a fraction of the total energy we presently use. Other traditional sources of energy, such as firewood, are still used by people living in technologically less advanced societies, but most people living in high-density urban societies use energy from oil, coal, and gas.

The amount of energy which is used per person in technologically advanced societies is many times more than that used by people in less developed communities. Our current urban societies depend on energy to support the billions of people now living in them. Cities all around the world are constructed in such a way that they are heavily dependent on energy to allow them to exist. If there was a shortage of energy, life in cities would very quickly become extremely difficult. Transport of food alone would present an immediate problem, as would heating in urban areas not located in tropical or subtropical climates.

The huge rise in the numbers of people living in cities since the early 19th century has in no small measure been helped by the increase in technology and use of energy. For example, Hong Kong doubled its population between 1950 and 1971 but increased its energy-use five times during that same period. This trend of increased energy-use per head of population is also taking place in the rest of China, as it is becoming more industrialized and urbanized, and the same is happening in other countries whose economies are growing.

As the way we currently live in cities depends so much on energy, it would seriously affect our current urban way of life if our present energy supply were to get scarce. In fact, without energy our current urban life, with its large populations, specialization, and mutual dependence, could not work. **The two main pillars of food and energy are essential in supporting the large populations of our complex and highly integrated urban organizations.**

**If anything happens to either of these two pillars, the whole social complex of a city could come crashing down, crushed under the weight of its large population. This risk is the price we pay for gaining efficiency through specialization.**

High population densities in cities or in rural communities have suffered from diseases, as they are easily transmitted from one person to another. As human populations grew during the course of history, disease epidemics have swept repeatedly through communities and caused the deaths of millions of people. It is estimated, for example, that during the 14th century, the Black Death reduced the population of Europe by between one third and a half. Throughout history, different diseases, sometimes collectively described as 'pestilence', seriously impacted on human populations in many locations in the world. For example, diseases periodically emerged in the growing populations of Asia and Europe, particularly at times when people were weakened by food shortages. Even though these diseases killed millions of people, other people did survive, and over time populations developed a certain level of resistance or immunity to them. People living in other parts of the world who had never been in contact with these diseases had no resistance to them. So when Europeans travelled outside Europe they carried these diseases with them and infected huge numbers of people with disastrous results. South African, Australian, and Pacific Island populations were decimated by diseases such as influenza, smallpox, measles, and dysentery. Following Columbus's arrival in America, the death rates from European diseases in some American-Indian populations are estimated to have been over 90%.

There are specific and general reasons for disease outbreaks in any living species. Higher population densities of any species bring increased risk of disease with them. As agriculture spread over the Middle East, Asia, and some time later also in Europe, the human populations increased dramatically. This fact alone increased the disease risk. Some reasons have been suggested why more dangerous diseases spread from Asia and Europe to the rest of the world rather than the other way around. Early agriculture in Asia and Europe had more types of domestic animals than in other regions. The suggestion is that most of our dangerous infectious diseases originally came from our farm animals. These diseases lived in animals that we domesticated, which then mutated slightly, as is usual, and some of these were then able to live in human beings. Humans lived in close contact with their farm animals, so a lot of cross-infection became possible. Growing human populations presented rich feeding grounds for these diseases new to humans.

Europeans did contract many lethal and debilitating diseases when travelling in the tropics and carried these diseases back home with them. However, most tropical diseases did not take hold in temperate climates in the way that temperate diseases did in the tropics, possibly due to the colder climate.

High population densities can influence the way we feel, the presence of many people around us can cause apprehension and stress. Many other animal species also react to high population densities, for example, they can change colour and behaviour as happens with locusts, pregnant females can reabsorb their foetuses as occurs in rabbits, and mothers can stop caring for their offspring or start behaving aggressively towards them, even killing them, like in rats and tree shrews. Stress-induced physiological responses are suspected of causing population crashes in deer living in high population densities. As mentioned before, it is thought that the reason why a number of animal species react to high population densities in this way is to avoid food shortages before they happen. It is seen as an inbuilt reaction shaped by evolution to maintain a steady balance between a population and its food supply.

Human beings also react to high population densities. We can become very aggressive towards each other, particularly when resources are scarce, as many violent episodes in human history have shown. High population densities also influence how many children we have. Average family size have decreased in many large populations even though reliable contraceptive methods were not available. Instances have been recorded of an increase in birth-rate as one moves away from cities into areas with lower population densities as we saw before. Circumstances are different for a person living in a small or in a large community, and it does affect the way we feel and the decisions we make.

As we know, since the introduction of agriculture our population has risen hugely, but the high population densities have in their turn also influenced the course of our ecology. Many of our inventions and particularly our recent innovations in technology could only have been accomplished in the context of high population densities, because division of work in urban life allows people to specialise in technology and inventions. The more people that can concentrate on specific problems and achieving inventions, the more answers we find and discoveries we make. Our technology has had a deep impact on our ecology. Our settled agricultural life and the specialization and coordination of large numbers of people in urban centres has made our technology possible. Certain types of research and development need a lot of coordination and resources to allow them to be carried out. One person or a small group of people may not have such resources, even if they could devote all of their time to it. Some of our current technology could only have been invented and can only be produced by large public organizations or large private companies. These organizations and these, often transnational, companies have the resources to carry out this type of work, and often have links with a network of other large organizations, so benefiting from other technological achievements and inventions. Such coordination of effort and investment of resources are usually part of, and made possible by, large centres of population.

The success and advantages of large human populations on the one hand, and the disadvantages and dangers of high population densities on the other hand, are another example of the way ecology consists of a balance which shifts between two unstable extremes.

If we had very low population densities, we would have to do without much of our technology, and the safety and quality of life that it gives us. If we all lived in extremely high population densities, the stress and conflict that this can cause, and the diseases that can arise, can lead to much suffering. Also, the lack of adaptability of big rigidly structured populations could result in large-scale tragedy if circumstances change too much. Therefore, the optimum balance to try to achieve a high quality of life lies somewhere in between the two extremes. This balance benefits from the specialization and inventions of large populations, but at the same time is not threatened by diseases, strife, stress, and resource shortages. In our recent history we have developed better health care and better understanding of social organization, and this has allowed us to move the balance more towards the high population density extreme. Previously, the optimum balance lay nearer to the lower density extreme as our knowledge and management of urban living was not as developed as it is now.

Our knowledge indicates that our evolutionary background has shaped us as social animals that lived in extended family groups of no more than approximately 50 people. In these small groups, individuals know each other personally. Our ecology has changed and now approximately half of the people on Earth live as part of large high-density populations in towns and cities. So this is a new ecological strategy for our species and we need to use our knowledge to manage and plan this new future. **While our specialized and mutually dependent urban life can be very efficient and has allowed us to attain many achievements, we have to be very careful in the way we plan our future, as urban living is only a relatively recent feature in the context of the overall history of the human species, and there are many problems associated with high-density populations.** For this reason The Natural Religion points out that the same specialization which gives us the efficiency of urban life and our astonishing technological advances also makes us very dependent on that system. If the supports of an urban system, such as food and energy, start to falter, a city can be crushed by the weight of its own population. The larger the population, the more pressure there is on its supports, and the greater the potential for large-scale disaster. Urban planning should therefore attempt to match expected population size to a conservative estimate of future supplies of essential resources.

The Natural Religion adds its voice to those who advise that the use of energy in urban centres should be reduced. This could be done in a number of ways, for example reduction of the daily transport of large numbers of people over long distances to and from work. One suggestion is the greater use of public transport, another is to design a city as a cluster of smaller centres. Such a network of connected 'village' or 'town' centres should have places of employment, residences, services, shops, schools, and sports and community facilities, ideally all within walking distance of each other. Another desirable aspect of such centres would be to have a social diversity within each village centre, different ages and socioeconomic levels should be mixed so that a much greater proportion of the people know each other personally. This is more like the situation in the small and medium-sized groups of which our evolutionary ancestors were part, and could be one way in which social divisions and the strife this can cause could be reduced. People who know each other personally tend to have a more balanced view of the other person than if they only know them as belonging to another social group. However, it is emphasized here that promoting social diversity within village centres is not a substitute for reducing the differences between rich and poor. Reducing poverty and raising the level of ownership in lower socioeconomic groups towards that of higher socioeconomic groups, is one aim of The Natural Religion in social and economic management all around the world.

While psychological stress due to living as part of dense populations does exist, the severity of problems such as disease, crime, and social division differ from city to city. Therefore we know that it is possible to increase the quality of life in a city with the right organization and management that reduces stress and avoids problems associated with high-density populations.

## 7.5: Limiting our populations

In Chapter 6 we saw that we cannot, as individuals, have and care for all the children we could potentially have, therefore, neither can we support the total worldwide population we are theoretically capable of producing. This means we have to limit our population. This is not surprising as every other living species on Earth either limits its own number, or its numbers are limited by the circumstances in which it lives. This is a basic characteristic of the entire biological system.

The relationship between a species and its resources is always at the core of the success or otherwise of that species. Amongst plant and animal species the situation does arise that one population could be overpopulated and suffering from a shortage of resources, while another population of the same species may have enough resources. Many species of plants and animals react to overpopulation in various ways, including limiting population numbers by reducing reproduction, but also migrating away from the resource shortages to areas with more resources. Similar to other animals, human beings also suffer resource shortages and overpopulation in one region and not in another. We also react by limiting our reproduction in various ways and by migrating away from the resource shortage to a region with plentiful resources.

The question however is: at what level should we limit our population? When will we have overpopulated Earth, or has that happened already? In November 2007 an estimated 6.63 billion people lived on Earth. Most of these have enough food for themselves and their children. Many areas produce more food than is locally eaten. However, some estimate that in recent years between 10 and 30 million people have died each year due to lack of food and over 500 million people are undernourished. Does this mean that we are now overpopulated?

In many instances during the course of human history, food shortages caused millions of deaths, but at the time of these famines there may have been enough food in the world as a whole to feed everyone. In fact, in some cases there was enough food quite close to the famine area, but those who were short of food did not have the money or other resources to buy or otherwise acquire this food. Poverty rather than food shortages has killed many famine victims. Does this mean that such an area was overpopulated, or is dire poverty the real problem, or is poverty a result of overpopulation?

In the very recent past, during the second half of the 20th century, famines have killed millions of people, while at the same time other regions in the world were overproducing food. This surplus food was being stored to prevent a collapse in food prices and possible bankruptcy of food producers. One rationale for such storage of food and economic control of food prices is, that large-scale bankruptcy of food producers would threaten future food supplies in that region. So, economic reasons can stop those who are starving to death from getting food which is being overproduced in another area.

One way to describe overpopulation, is to say that it occurs when there is a shortage of resources, but this does not take into account when there is actually enough food to feed everyone except for those people who are too poor to buy it. Another aspect of overpopulation is that large populations can wear out, and use up, a resource over a number of years, decades, or centuries. Soil degradation can take decades, and ground water reservoirs can also take years to exhaust. Should we speak of overpopulation during the period that the particular resource is decreasing but not used up yet? There are resources which we are presently using, which one day will be used up, such as fossil fuels and some underground water reserves. Does this mean that ever since we started using fossil fuels or any other of our finite resources that we are overpopulated?

We could say that we are overpopulated in a certain area if there is a shortage of resources. So, we could have quite a low population yet be over-populated if there are not enough resources and people are poor. On the other hand, in another region with a larger human population, we may not be overpopulated because there are enough resources and people are richer. Does this mean that only poor communities need to limit their populations?

Richer communities tend to use more resources than poor communities because they have more. They can afford to use more resources per head of population and can also be more wasteful with resources. In other words, richer communities can use up supplies of certain resources faster than poorer communities because they have the wealth to do so. This demand can also raise the cost of resources beyond the reach of some poor communities. Should it, therefore, be richer rather than poorer communities that should limit their population?

During the course of recent human ecology we have increased our resources in a series of waves. Each time we found a way to increase our resources, our world population also increased. In our history we have experienced many regional shortages and millions of people have died from famine, but on a worldwide basis we have, over the centuries, managed to create new and higher levels of resources many times over.

The first indications of an increase in the world's human population appear to have been associated with the development of better tools and implements associated with our early hunting and gathering lifestyle. The next big increase in resources was due to the advent of agriculture. A series of new innovations in agriculture since then have more or less continuously increased our agricultural food production. Hand in hand with agriculture, our development of industrialization gave our species further scope to increase in numbers. Another wave of unprecedented population growth in the twentieth century went hand in hand with the revolution in innovation and technology which also happened then. If we can continue to create new ways to increase production of our resources, we could theoretically also continue to increase the number of people sharing Earth. However, ways of creating new resources will most likely slow down and stop at some stage because our resources are not limitless. If we deplete our essential resources and continue to have a large world population, famine, violence, and diseases will cause even bigger disasters than those that have already taken place. This is one way in which our population can be limited, but it is not the only way.

Alternatively, we can try to calculate how much resources we are likely to have in the future and judge how much our human population will need. Then we can match our future population to what our resources are likely to be, by using population planning methods which do not cause turmoil and suffering. Safe contraceptive methods, informed and unrestricted family planning, knowledge about the age profile of our populations and demographic planning, can all be part of limiting our populations in a peaceful and humane manner. These are the methods we should use to limit populations so that they match our resources and maximize the quality of life of people all around the world. If we don't limit our populations in humane ways, we can either limit them in more oppressive ways or a lack of essential resources will result in a situation which is out of control and results in death, suffering, and a catastrophic collapse in population.

To identify exactly what population level will optimize our quality of life and make use of our resources in a sustainable way is a difficult task. Cultural differences, complications in economic management, conflict over control of resources, and varying political and scientific opinions all contribute to making it difficult to identify the right population level. There are, of course, further problems associated with actually achieving this population level subsequently, but all over the world the same rule applies: we need resources to live. If we allow our populations to get out of balance with our resources we are faced with disaster. This book accepts the principle of limiting our populations all around the world so that we do not run out of essential resources, but it is stressed in The Natural Religion that the manner in which this is done is all important and should result in raising and maintaining the quality of human life on a worldwide basis, not reducing it. This means that poorer regions in the world are made richer, and that resources are made as individually available in these regions as they are in richer regions. It also means that the use of resources in all regions of the world should be evened out, and wasteful use of resources eliminated. We should also put full effort into researching and developing ways to substitute sustainable resources for those resources that are currently non-sustainable. Regional self-reliance should be promoted as much as possible of course, but abundances and shortages of resources will always occur in different regions and at different times. Therefore, resources have got to be spread evenly amongst human communities around the world. Our worldwide economic system transports and exchanges resources around the world, but this same economic system should not be the cause of famine, suffering, and death because some are too poor to buy food.

**One ambition of The Natural Religion is, therefore, that economic mechanisms be developed by which food can be supplied worldwide to those who are suffering from famine that also protect the economic viability of other communities. People dying from lack of food in one part of the world, while there is a surplus of food elsewhere, is an utterly obscene situation and completely unacceptable.**

**Amongst the conclusions that together form The Natural Religion, is that we should balance the number of people sharing Earth with our resources, which means that because of the recent unprecedented growth in our human population, together with the fact that many of our resources are finite, we have to limit our world population in a fully informed and free way, using safe contraceptive methods and with people having full control over their own family planning.** If we don't achieve this, limitation of our populations will be imposed on us in other and likely undesirable ways.

As we saw, our economic system plays a crucial role in the quality of life of people worldwide. Therefore, the next chapter looks at the role of our economy in our lives.

# Chapter 8

# Distributing Our Resources

> ＊ Life needs resources and so do we;
> we share resources but we also compete for them ＊

## 8.1: Our resources fuel our life

How we distribute and share our resources – more commonly called our economic system – involves balances between two extremes, like other aspects of life. Traditionally religions have dealt with this fundamental part of human existence in ways ranging from religious ceremonies for agricultural events and regulating the sharing of food, seed, and water; to condemning covetousness and usury.

During the time that we evolved as hunter-gatherers we lived in groups. The members of these groups helped each other. One person gathered roots and tubers from the ground, while another person went to a different location where fruits were more easily found. A third person hunted for animals and brought meat home, and a fourth person caught fish. These people then shared their roots, tubers, fruits, meat, and fish. It is much more efficient if people cooperate with gathering and exchange this range of foods, rather than if each individual person travels to all the different places to gather the various foods every time he or she needs to eat. Our evolutionary ancestors, therefore, distributed and shared their resources within their group.

Humans need other resources besides food. As hunter-gatherers we also needed tools and implements, as well as clothes and shelter as we moved to colder climates. The greater the variety of resources we used, the less likely it was that each person could find or make everything they needed. Certain foods, clothes, or implements for cooking and hunting may not be available in all areas. For example, a population of fishermen living near the sea could have collected dried fish, sealskins, fish oil, and salt. Hunters living more inland could have furs, wood, forest foods, and objects made from stone, ore or clay found in upland regions. Hunters and fishermen would typically exchange these resources and the products they made from them. So, from very early on in our development as a species, each person relied on others to supply them with at least some of the resources they needed, however a person would have to give something in return for what they received. In fact, whatever a person receives in return has got to be in proportion to what they give away. If one person gives away ten animal furs which took a week to catch, and receives a small sack of berries which only took a day to collect, he or she would be better off to gather the berries themselves and also have seven or eight furs. This balancing in terms of value and effort it took to acquire the traded items is essential for the survival of the people involved in the trading. We need a certain amount of resources to live, and if we give them away and get little or nothing in return we will end up with too few resources to survive. Therefore when trading, whether as part of a basic bartering system or as part of our present-day complex economic system, we have got to balance the value of the resources which we exchange. If we don't, we compromise our survival.

If the food supply in an area is not dependable, then we have to move to another area to find more food. It is because dependability of supply of essential resources is so important in our human ecology that we developed the concept of ownership. Ownership means that a person or group of people have control over a resource that they know will be theirs to use in the future and that no one else can use. If we care for and develop something we own, we know that we, and not someone else, will benefit from it in the future. So we can afford to devote time and energy to something we own because it is an investment in our own future, we will benefit from the invested time and energy.

Ownership is part of the ecology of other animal species besides humans. Many species have territories which belong to individuals or groups of animals. These territories are often fiercely defended against others, and with good reason, as the survival of the individual or group depends on the resources within that territory. In fact, if a group gets too big for the resources contained in a territory some individuals may be expelled from the territory by the rest of the group. Animal species ranging from insects to birds, fish, and mammals defend territories. Our domestic dogs protect our homes because it is also their home.

This book recognizes that the reality of human life is that, like all biological life, we need resources to live. We need a range of different resources, and most human beings rely on others to give them at least some of these. However, realistically, we cannot afford to give everything away without receiving something in return, we need to exchange resources. **Therefore, the conclusion about sharing and ownership in The Natural Religion is, that ownership is an important means to an end, the end being predictability of resources.** Ownership is part of our ecology as it is of other animals' ecology.

## 8.2: Sharing versus ownership

Planet Earth's resources allow us humans to live, and we have to share these with each other because all human beings need them. Not to share resources threatens people's existence. At the same time, ownership makes supply of resources predictable, and as owners we can afford to invest time and energy into enhancing and caring for our resources. So, on the one hand, we have to share resources with others, and on the other hand, we have ownership to make resources more reliable but which restricts availability of resources to their owners. This is another situation in our ecology where the two extremes, of sharing everything, and of outright ownership and not sharing anything, are unstable and undesirable. What actually happens, and has developed during our evolutionary history, is a balance that moves between these two extremes.

Sharing everything would mean that we have free and uncontrolled access to all resources by every person on Earth, and planning for the future would not be possible. Resources would be unpredictable and those who were strong, or lucky enough to get there first, would use the resource while others would get little or nothing. Even those who got to a resource first, could not afford to share it with others even if they wanted to, because they could not be sure that they would get a share of other resources in return. In a free-for-all it is risky to share resources, as future supply is uncertain for everyone. The only practical thing to do in a free-for-all is to get what you can, while you can, because if you don't someone else will. In a free-for-all no one can be sure that the resource will be there in the future. No one can plan ahead if resources are unreliable, and life itself, therefore, becomes unpredictable.

The other extreme is not to share anything and to have strict ownership of all resources, and not allow people access to any resource they don't own. This can mean that, if people do not own a resources which produces food and if those people also don't have money to buy food, they simply die. If ownership is the main priority, life becomes very harsh and uncompromising. As mentioned in Chapter 7, millions of people during the course of human history have died because they were too poor to buy food. **We, as a species, share planet Earth, and our whole species have to share our planet's resources to survive. People should not be allowed to die from hunger because our ownership system is too strict to allow existing food to reach them. Ownership should not kill people.**

Human society is a very social and integrated society, what happens in one part of a community affects other parts. So, if we make a society harsh and uncompromising towards its poorer sections, this will inevitably also influence the rest of society. For instance, if people face death even though the resources exist that could save them, they will try to take these resources by force because they have nothing to lose. Those that have resources will react to protect themselves, possibly with even greater violence, resulting in a violence-ridden society for everyone. In addition, resources are often wastefully destroyed during the course of violent conflict. In fact, an unyielding system of ownership does not even have to be so extreme as to threaten lives, to result in serious tensions in a society. These situations have, unfortunately, occurred many times in human history and continue to do so.

In most societies throughout human history sharing resources has been combined with a certain level of ownership. Sharing the resources which planet Earth has to offer has been part of human ecology since our emergence as a species, but certain levels of ownership have also developed. The balance between the two extremes benefits from the predictability of ownership as well as from everyone getting what they need. Sharing resources is currently part of hunter-gatherer life – at the end of each day whatever is brought back to the camp is shared between all the members of the group. We therefore presume that our evolutionary hunter-gatherer ancestors also shared their resources, and whatever was brought back was not seen as solely being owned by the person who caught the prey or gathered the plant food. Yet, the group as a whole may very likely not have welcomed sharing their territory with other groups, particularly if resources in their territory were not all that abundant. Thus, a balance could be struck between sharing within the group and ownership between groups.

Irrigation water used in early agriculture was also shared. No one person could divert all the water to their own plot of land. All the farmers in an area shared the water, so it was carefully divided and each received a regular and reliable supply to irrigate their crops, but all farmers together took care of the system of irrigation ditches and limited the water to their group. In most societies the living space that each person occupies is also carefully divided. Part of it belongs to a person or family, and part is shared with other members of the community. A hut or house is usually a private area reserved for a person or members of a family. However, the space connecting private huts or houses together is typically shared by the community. This is still the balance which many traditional as well as most modern societies strike between owning one's own private space and sharing communal space.

The age-old system of collecting taxes is also part of the balance between ownership and sharing resources in a community. Taxes are normally used to build and maintain structures and services that a whole community uses and shares. The proportion of our personal resources that we pay in taxes, is the balance which is struck between ownership and sharing.

All animal species, whether they are very social or not, divide and share their resources, and also occupy territories or personal space from which others are excluded. But, territories usually have their limits and outside these boundaries, space can be shared by the community. Animals keep a certain amount of food for themselves, but also share food with others, the amount of food and the time when it is shared, or the individuals with whom it is shared, is the balance which is struck between owning and sharing in each particular case.

**The Natural Religion's position is that the resources on planet Earth are needed by all biological life, but ownership is part of the way in which animals, including ourselves, care for and manage resources in order to make their supply predictable for the future. Striking the balance between sharing and ownership is crucial to our welfare and quality of life.**

## 8.3: Wealth and poverty

In historical times hunter-gatherers tended not to accumulate and store stocks of food and other resources. So also, present-day hunter-gatherers will stop gathering when they have enough for their immediate needs, even though more food and resources are available in their territory. As hunter-gatherers live in low population densities they know that ungathered and unused resources in their territory will remain available to them at a later date. The territory in which a hunter-gatherer group lives is effectively their food store.

When we started to practise agriculture, crops could only be harvested at certain times of the year. This meant that we had to store this harvest, so we had food during the time when we had no harvest. We could not leave it in the field as the crop might rot, and as well we had to prepare the field for the next crop. We also kept domesticated animals that we would kill if and when we needed them. Agriculture is more efficient in producing food than hunting and gathering, so that farmers can produce much more food than they and their families need, which is then exchanged for other resources. This means that a prehistoric farmer had a surplus, which he or she had to store, and could exchange for such essentials as clothes, fuel, cooking pots, farm implements, building materials, and weapons; as well as such non-essentials as jewellery, pigment for colouring clothes, and other fashion items. Wealth and poverty have existed side by side in communities the world over, ever since we started practising agriculture. Present-day human ecology is primarily dependent on agriculture for food production, and trading our surpluses has become an intrinsic part of the way we gain our resources.

To trade one resource for another in a bartering system, each person needs to have a surplus of what the other needs. Thus, there can be a problem if people have surpluses and shortages which don't match, for example if one person has a surplus of clothes and another has a surplus of fuel but both of them only need food they cannot barter. To solve the problem of complicated rounds of bartering involving more than two people, we have developed universal bartering media. Special shells or large stones have been used for this in the past, but as briefly touched on in Chapter 4, metal coins and paper money have survived the test of time. A universal bartering medium has a set value against most resources, which is generally accepted by everyone in communities and societies. Everyone agrees that a certain amount of money represents a specific amount of food or other real resources. Trading using money means that only two people need to be involved and both can exchange the same money for other resources they may need.

A further advantage to a universal bartering medium such as money is that unlike, for example, foods, it does not rot and become unusable over time. A person can accumulate money and exchange it for resources at a future date, so having a lot of money is also accepted as wealth, as it represents real resources. One aspect of money is that a group of people or a society has to agree on how much money is equal to how much real resources. We cannot eat money, nor does it keep us warm, physically protect us, or cure us if we are sick. Its real value is that people all agree that it represents resources. However, money can change its value. Gradually over time more money can be needed for a certain amount of food and other resources, this is inflation. In addition, great social upheaval, such as war, can cause a certain type of money, or currency, to lose its value completely. People then will not exchange any resources for that currency anymore and it loses its use.

It is when social agreement on the value of a currency breaks down, that people fully realize that money is only a substitute for resources and not a real resource. **While money makes it easier to distribute and exchange resources rather than a complicated system of bartering real resources,**

**money is valueless if we don't agree on how much real resources it represents.** For example, on one occasion money lost all its value, and seriously devalued on a second occasion, in Germany during the first half of the twentieth century. There are many other examples of money losing its value in the wake of a war. The Natural Religion therefore warns that if for any reason the common agreement on the value of money stops, real resources are then only those which can feed us, keep us warm, protect us, and in general help us to survive.

It is in the interest of individuals to gather resources, and people who succeed in gathering many resources are wealthy. Wealth, in biological terms, means having more resources than is needed to care for one's own health and welfare, and to reproduce and care for one's offspring. Wealth makes predicting that we will have enough to live on more certain, and it is therefore a normal urge for people to want to accumulate wealth of one form or another. In agriculture, wealth can mean having a lot of land which can produce food. Wealth can also mean having surpluses of other resources such as clothes, fuel, or implements of all sorts, and wealth in most regions of the world also means having a lot of money. In most communities everyone does not have the same amount of wealth. Some people accumulate more wealth than other people do and, as with any living species, human beings compete for resources and some are better able to compete than others.

Poverty means that a person does not have enough resources to care for one's own health and welfare, and to reproduce and care for one's offspring. In most human communities around the world, some people are wealthy but some are also poor. At present our economy has resulted in large differences between rich and poor in the world. An estimate in 1994 reported that the poorest 45% of the world's human population (then over 2.4 billion people) lived on only 4% of our at that time total world Gross National Product. This was approximately the same amount of wealth as owned by 358 of the world's richest billionaires. 25% of our world population have been designated as living in poverty. 800 million people are classed as chronically hungry, and in recent years each year between 10 and 30 million people are estimated to have died from the effects of lack of food, this included 11 children per minute in 2003.

It is also estimated that the richest 20% of the world's population owned 80% of the world's wealth. This inequality in wealth and ownership is bad enough; but what makes it worse is that the level of inequality is currently reported to be increasing rather than decreasing. For instance, between 1960 and 1991 the difference between the richest 20% and the poorest 20% in the world is reported to have increased from 30 times to over 60 times.

The foreign national debt of countries is reported to have increased from $650 billion in 1980, to $1.3 trillion in 1990, $2.2 trillion in 1997, and $2.8 trillion in 2005. This represents approximately 45% of the Gross National Product of developing countries. To put these amounts of money in context, the foreign national debt of the 20 poorest countries in the first years of the twenty-first century could have been ended with $5.5 billion, which is similar to the amount it cost to build the holiday theme park Euro Disney near Paris. In Europe people spend 7 billion euros per year on perfume, which could supply 2.6 billion people with clean healthy drinking water.

Poor people tend to be less healthy, have less education, and have fewer life options available to them and these trends are seen both within countries and internationally. Particularly when there are high income inequalities in a region, you find more people in prison, more violent crime and murders, more unemployment, worse school performance, more policing, and a generally worse economy. One instance of the influence of the range in incomes on a society is the example of Japan, which has one of the smallest ranges in incomes and also has a very low crime rate and greatest longevity compared to other countries.

Poverty affects people deeply through bad health, short life expectancy, and the prospect of even more suffering and deaths if conditions worsen. **In fact, the poorer you are the closer you are to staring death in the face. Poverty threatens survival. Lack of education contributes to lack of choices in life, all-round poor quality of life, and poor health. An economic system which has substantial numbers of poor people in it, functions at a lower level than it would if these people were richer. The first task of our economy is to share and distribute resources, and the multiple of millions of poor people are proof that it is not effective in doing this. We humans should work towards excluding poverty and the suffering it brings with it from our ecology.**

Apart from the serious effects which poverty has on those who are poor, it also affects those in that society who are not poor. These effects directly impinge on richer people, and exist in addition to the compassion which richer people should feel for the suffering of their poorer fellow human beings. In a country that has very few poor people, a high percentage of the population is economically active; and nearly all people produce products or services in their lives, and are also consumers of these products and services. Thus, the economy of this country has reached its full potential. However, if the same country has a significant proportion of poor people, these cannot be as economically active as their richer compatriots. Poorer people have fewer resources and tend not to have as much education as richer people, and are therefore limited in the products and services they can produce. Poorer people also have much less money to spend on products and services. The combination of these two factors reduces the size of that country's economy to well below its full potential. In a country which has an economy which is operating below its full potential all inhabitants, rich and poor, are disadvantaged by this.

Big differences in incomes can also bring about violent conflict. If poor people do not see any prospect for themselves or their children of ever becoming wealthier, it can result in a violent reaction to this state of affairs. When people think they have nothing to lose, they feel justified in adopting high risk strategies such as violent conflict as they are fighting for their very survival. Richer people will then fight back to protect themselves and what they own, and a lot of their greater resources will be put into a violent response. Such a situation creates a society in which a lot of time and resources are dedicated to killing and destroying – a society in which fear, hate, and suspicion are the predominant feelings, rather than cooperation, friendliness, and trust. The whole society, rich as well as poor, will then have a quality of life very much below what it would be if there was no poverty. The extent to which violent conflict is now part of our human ecology and economy is indicated by the various reports that rank the arms industry amongst the three to four biggest industries in the world, together with the oil industry, health care, service industry, tourism, gambling and the illegal drugs trade. **Poverty not only reduces the quality of life for those who are poor, but also reduces the quality of life for all of those who live in that society.**

Reduction of the level of poverty, and the narrowing of the gap between rich and poor, are essential parts of the future management of our human species if we want to have a reliable world economy and a good quality of life. The presence of poverty in our world population causes suffering for those who are poor and also adversely affects those who are richer by restricting the world economy. Violent conflict, which routinely results from tensions between rich and poor, will continue to threaten both rich and poor human populations. The way in which the root-cause of these dangerous tensions can be taken away is to close the gap between rich and poor by making the poor richer.

**Poor people have less economic and political power than richer people. So richer people have to initiate most of the changes in the way we manage our economies. Richer people have to make poorer people richer and make sure that the first priority of our economy is to share and distribute resources, because they have the economic power to do this.**

Part of The Natural Religion is the viewpoint that the primary function of our world economy is to share and distribute resources, and therefore the poverty which presently exists in our world population is seen as proof that our economy is failing in its primary function. It is estimated that in recent years severe poverty caused the death of tens of millions of people every year, while the amount of money which would alleviate this poverty is reported to be as little as approximately 1% of our global income. **One of the main aims of this book is to help improve the quality of life for all people, and the alleviation of poverty is one of the main ways in which to achieve this.**

Those of us who are richer are reminded, that poverty in global, as well as local, society affects all of society both rich and poor. Apart from the compassion which richer people should feel for poorer people, richer people are also materially adversely influenced by poverty in their community by a reduced overall economy and sometimes also by violent conflict. Both these factors seriously reduce the quality of life in society as a whole, not just for the poorer sections; and if the difference between rich and poor in our global economy is allowed to get greater, then the effects of poverty will get worse. **We do have the potential to live in more egalitarian societies because we did so in the past, and therefore it is part of The Natural Religion that it is realistic to conclude that we can adapt our current economic system to reduce and eliminate differences between rich and poor in all our societies.**

## 8.4: Balancing our economy

Most countries are directly connected with our worldwide economic network. Some more remote groups of people may not be directly connected to it, but are usually still indirectly influenced by it in various ways. As we have already seen, our economy fulfils a central role in our human ecology. With the recent unprecedented increase in the number of humans on Earth our economy has become ever more complex and ever more crucial as more than 6.6 billion people depend on it for their survival.

How does our economy solve the conflict between ownership and sharing our resources? What is the best way to strike the optimum balance between these two extremes? How can we bring back a greater equality in wealth and status into our modern agriculture based societies? How can we control fluctuations in our economy and stop damaging recessions? How do we manage our economy in such a way as to share and distribute our resources more evenly amongst all humans?

During the course of our human social and economic evolution we have tried different ways of running our economies. As we saw, our hunter-gatherer ancestors had an economy in which ownership was communal rather than individual, and large differences in rich and poor did not exist. Then when we started to practise agriculture, ownership and greater differences between rich and poor developed. During the feudal system, ownership became strict and social position gained great importance and became dependent on wealth. Ownership continued to be strict during the industrial revolution, and social classes were still very distinct and people reacted to this, giving rise to the socialist and communist movements. These movements advocated the redistribution of wealth and abolition of socioeconomic classes, by taking major resources out of private ownership and putting them into communal or public ownership. In the countries in Europe where this type of economy was tried, people appear not to have worked very productively. This has been interpreted as being largely due to people not seeing themselves as part of communal ownership, whether people put more or less effort into their work made little difference in the return they got out of their endeavours. The connection which people felt between personal effort and return seems to have been too remote or non-existent.

Another problem with these economies which were owned and run by the state is that they tended to be controlled by a bureaucracy made up of a small number of public officials. Because of the centralized and non-democratic nature of the societies in which communist economies existed, these officials could not effectively be criticized or challenged in any way by the citizens. This dictatorial system of managing the economy led to inefficiencies and mistakes that were very difficult to correct. A lack of ability to identify and acknowledge mistakes, and therefore correct them, is a common fault of many hierarchical systems.

Another type of economic system which arose was the market economy in which the forces of supply and demand were given free rein, and which is firmly based on private ownership of resources. In this type of economy, the laws of supply and demand determine prices for resources, products, or services and the system encourages and rewards personal effort and achievement. While this system seems reasonable and fair in the way it operates, it does not in fact result in a reasonable and fair society if it is allowed to develop uncontrolled. The reason for this is that the job of each person who works in the private economic sector is to make themselves and their company an economic success. This means getting as much profit as they can for their product or service. This, however, is not in the interest of the public in general. The private economy is about ownership, and is not oriented towards sharing. Being based on ownership, and because resources are usually limited, it also means that different people and organizations in the private economy compete with each other for economic success. So, the job of a person working in the private sector is to be economically successful and out-compete their economic rivals. If this system is allowed to progress without a certain level of control, the larger and more powerful economic organizations will out-compete all the smaller ones and resources too may be wasted. This means that the whole economic system ends up being controlled by a handful of large and powerful monopolies.

Economic monopolies are not in the interest of the public at large as a monopoly looks after its own organization, not society in general. Economic monopolies, like any economic organization, will try to get as high a price for its products as possible, but because it is a monopoly it has little or no competition and can demand what it likes by controlling the supply of its product. This is at the ownership extreme and very much away from the sharing extreme, and can give rise to a harsh and uncompromising economy in which the poor can find it very difficult to get the resources they need. In an economy dominated by monopolies, a small number of people end up controlling the entire economy. This, ironically, was also the outcome of many communist economies, although the two types of economy started from very different ideological positions. As with a non-democratic state-controlled economy, it is very difficult to challenge or effectively criticize the people controlling a monopoly economy, and this again means that mistakes are difficult to acknowledge and correct. Economic monopolies are also inclined to suffer from strict hierarchies that function as dictatorships, with all the faults thereof.

As we have seen, poverty in a society causes the poor to suffer, limits the economy to a level below its potential, and reduces the quality of life of all in that society. To reduce the difference between low and high incomes also reduces crime and improves health, employment, education, and promotes a more active and buoyant economy. So the economy as a whole should be managed in such a way that it can fulfill its role of distributing and sharing our human resources, and to do this we need a balance between ownership and sharing. And that means that while we need the efficiency and dependability of privately owned economic units, we also need to stop the development of monopolies.

Public authorities can regulate the private economy in such a way that it can operate successfully, but at the same time does not develop into monopolies which smother economic diversity and competition. The first priority and main concern of public authorities is to protect the public and society in general, so this job cannot be done by any section of the private economy as their first priority has to be to ensure the success of their own economic unit. What is essential in the way public authorities function is that they are not corrupt or influenced by the economic power of private enterprise. Public authorities have to be independent of private commerce so that they are completely free to regulate and control in order that society as a whole, ie: the general public as well as the private economy, benefits.

Reducing poverty and equalizing levels of wealth, as well as having many sociological advantages and improving general quality of life, can also raise the level of economic activity. The private economy needs customers and it can do more business with richer people than poorer people, so it is also in their interest to be part of a society which has little or no poverty.

Public authorities also have to care for the private economic sector. Without the products and employment which that sector produces, the economy would operate well below its potential. Private enterprise creates efficiency in producing goods and services, and if public authorities impose too many restrictions the private economy will suffer. The function of public authorities is to create a society with a good quality of life, and a healthy and vigorous economy is part of this. The private sector may have to be helped and supported so it can be economically successful. To achieve this, **public authorities have to strike a balance between protecting the interests of the private economy and protecting the interests of society at large. This includes creating the conditions which will result in an optimum diversity in the economic sector, including small, medium, and large economic operators.**

Different-sized economic enterprises have different advantages and disadvantages. Certain products, particularly those involving a high level of technology, can possible only be developed and produced by large companies, but most successful farming operations tend to be family-run businesses. Large organizations have disadvantages which smaller enterprises don't have and vice versa. Economic conditions that favour one type of economic enterprise over another inevitably change, after which previously disadvantaged operations can become the more successful performers. And different types of products and services are most efficiently produced by different types and sizes of economic enterprise. Diversity within the private economic sector creates more economic adaptability and possibilities, and therefore greater economic stability in the medium and long term.

It has been the experience all round the world that economies go through cycles. Times of strong economic growth have been followed by economic recessions. Individual economic enterprises also go through periods of strong economic activity followed by a slow down. Most economic operations have to change what they produce or the way they operate as time goes by or they will become less successful. Economic conditions, as with all ecological circumstances, change over time. Diversity and adaptability should be built into an economy so it can absorb these changes and even benefit from them.

Generally though, those who are involved in the private economic sector need to focus on short-term success and strive for economic growth, but simple economic growth cannot last. To base future economic management on attempting to have constant economic growth is shortsighted, particularly where the human population is not increasing. Economic management policy should accept that there will always be constant change in economic conditions, and in order to avoid damaging recessions in economic activity, a capacity for continual change should be part of the way our economies operate. While one economic sector is experiencing growth, another sector should be changing so that it is ready for its growth phase when growth in the first sector slows down. Economic growth should also be controlled so that it does not happen too fast because this could bring about a sudden uncontrolled recession. Those in the private economic sector need to primarily focus on the day-to-day running of an enterprise, but part of the role of our public authorities is to take a longer term view and plan the health of the economy over a longer time period. This is in the interest of the general public as well as the private economic sector.

Ideally our economy's most important function is to share and distribute our resources fairly and efficiently to human beings all over the world. Wealth can be accumulated in the course of economic activity and is needed as a prerequisite for many of our more complex economic activities and other achievements, but our economy should share and distribute resources and it should use accumulation of wealth as a way to achieve this. It should not be the other way around, in other words, where our economy's main aim is to accumulate wealth and use the distribution of our resources as a means to achieve this. It is our public authorities' function to strike the balance between sharing and how much wealth should be accumulated, and part of this is to balance our economies in terms of size and type of economic enterprises. This balance may have to shift towards one extreme or the other depending on circumstances, as with the balances in all other aspects of our ecology. However, this economic balance needs to be struck by public authorities rather than by the private economic sector. Each person's first priority in private enterprise is to ensure the survival of that particular economic enterprise; it is not to take care of society as a whole. To care for and shape society is the role of our public authorities.

One very important criterion which can be used to judge whether the balance that is struck between ownership and sharing resources is correct, is the presence of poverty. At the present time in some countries poverty has been reduced compared to the poverty which existed in recent historical times.

However on a worldwide basis, poverty is widespread and the gap between rich and poor is increasing. As mentioned already, this can be taken as proof that the balance between ownership of resources and sharing resources in our world economy is currently far too much towards the extreme of ownership.

**That the role of the private sector is vital in our world economy, and that it should be actively nurtured when necessary is fully recognized by The Natural Religion; and because the function of those involved in the private economy is to care for and promote their own particular enterprise, it advises that public authorities control and manage the private economic sector in such a way as to benefit society as a whole.** One aim of public management of private enterprise should be to create an optimal balance of economic enterprises within the private sector that serves the general public and creates an adaptable and stable economy in the medium and long-term.

## 8.5: A sustainable, stable, and diverse economy

Many of the resources which fuel our economy have limits. For instance, we only have a limited supply of fossil fuels: oil, coal, and gas. Some of our resources, like clean drinking water do replenish themselves but in too many instances we are using them at a faster rate than they are being replenished. In both cases of fossil fuels and drinking water we cannot continue to use them as we are now, and our economies which are based on these resources are therefore not sustainable. Food production is essential to our survival. As our population increases, our food production also has to increase; and during the 20th century this was by a massive 400%, as we went from 1.5 billion to 6 billion people. Therefore our demand for food also experienced unprecedented growth and, although many regions around the world did experience famines during the 1900's, food production helped by advances in agricultural technology did increase massively. The question is, however, whether these increases in food production are sustainable. For example, one of the biggest users of water on a worldwide basis is agriculture. In certain regions where food crops depend on irrigation by water pumped up from deep wells, the water bodies from which this fresh usable water is being pumped are filling up at a slower rate than we are emptying them. Many of these fresh water bodies are therefore shrinking, and so this type of food production is not sustainable.

Another aspect of food production is that unless nutrients and other essential soil constituents are being replaced, soils will degrade. Soils can become infertile or be eroded by wind and water due to incorrect soil management, and if we have no fertile soils we cannot produce food. **Food is a resource**

**which is basic to our survival, it is the fundamental part of any economy. If food production is not sustainable then the whole economy is not sustainable.**

Soil nutrients can be replaced, other constituents of fertile soils need to be protected and enhanced and we can treat our soils in ways which counteract erosion. In countries with low rainfall we can irrigate in ways which avoids salts that make the soil infertile from accumulating in the surface layer. So, first of all, it is important that the need for sustainable agricultural practices is communicated to all, including to the people who work on the land; but what is equally important is that those who do farm sustainably are not economically disadvantaged because of it. If two farmers produce grain and one replaces nutrients and practises long-term soil care and the other does not, the farmer who treats the soil more sustainably will have to charge more for the grain than the farmer who just grew the grain and did not give any resources back to the soil. For example, the price charged for crops produced using irrigation normally includes the cost of irrigation, but usually does not include the cost of replacing the water in the ground. Irrigation is not sustainable unless the water it uses is replaced at the same rate as it is being used.

In the world economy, competition drives down the cost of products and that is generally good, but if unsustainable practices are used to cut costs then the economy itself will become unsustainable. If we ignore certain costs like replacing irrigation water, then we are just postponing paying the full and real price for our product, we are in fact paying an unsustainable price. It is in these situations that public authorities should intervene in economic management. The economic situation should be such that on the one hand competition keeps prices at a fair level, but that on the other hand sustainable production of food and other resources is rewarded economically. This means that sustainable production should be rewarded more than unsustainable production, as this is in the interest of both the consumer and the producer – in fact, the whole economic system. Certain costs need to be paid by society as a whole because it is important that land produces enough food for people to eat, no matter who that land belongs to. It is in the interest of society that enough farmers produce food to feed everyone. For this reason, farmers need to be rewarded realistically for the food they produce, and they should not be forced into unsustainable practices by demand for low prices. Society as a whole needs to create the economic circumstances in which food can be produced sustainably.

**The view included in The Natural Religion is that food producers should be rewarded and not economically penalized for using sustainable food production methods. The full cost of sustainable food production should be borne by all of us as we all need food to survive.**

Limited fossil fuels also present a serious threat to the economy and survival of many millions of people. The reason for this is the nature of urban and industrialized societies. As we saw in Chapter 7, urban life is very dependent on energy, most of which comes from fossil fuels. Food has got to be transported to millions of people from outside cities where it is produced. Within cities, millions of people are transported to and from their homes every day. Many cities are located in temperate and colder climates and there we need heating to keep us warm. All this takes energy. The whole urban economy relies on energy in order to work. Presently around half the world's human population lives in cities. This means that billions of people are completely dependent on energy for their day to day survival and, as we already discussed, if energy supplies to big urban areas were to suddenly stop, it would be calamitous. Fossil fuels are limited and if we continue to use them they will run out one day. This means that the energy-dependent urban economy and way of life will lose its main source of energy. **Therefore the present fossil fuel based urban economy is unsustainable unless another equally large, but this time sustainable, source of energy is discovered.**

Sources of sustainable energy do exist. Sunlight, wind and wave power, and heat from volcanoes are sustainable, and so is the flow of water in rivers fed by sufficient rainfall and with a sufficient fall in height. Certain plants grown in a sustainable manner can also provide sustainable energy. However, less than a quarter of the world's energy supplies are from sustainable sources and this is even less in industrialized and densely populated parts of the world where we rely proportionally more on non-sustainable fossil fuels for energy.

Supplies of oil and gas are smaller than the world's coal supplies. However, burning coal releases more harmful pollutants into the atmosphere than the other fossil fuels, thus oil and gas are preferred, being cleaner fuels compared to coal. Coal is also finite, so would merely be a stop-gap and non-sustainable. Nuclear power was at one stage thought to be a cheap source of almost unlimited energy. While this is potentially still the case for nuclear fusion reactions (in effect, producing energy the same way that the Sun does), energy produced by nuclear fission reactors is non-sustainable as its fuel, like fossil fuels, is limited, and has also proved to be expensive. In addition, nuclear fission reactors can also pose health threats for people working in that industry, as well as for the general population, from the wastes which are produced by the normal nuclear fission process as well as accidents such as explosions. The reason for this is that the nuclear material which is released gives off lethal radiation and continues to give off this radiation for hundreds of years. Low doses of nuclear radiation can affect a person's health, and larger doses can kill immediately or over a period of time. Nuclear power stations are usually designed to operate for a certain period of time only, after which it is intended that they are closed down. However, when they are closed down they have to be sealed for many decades.

Apart from the fact that nuclear fission fuel is finite, a big problem with nuclear fission is that the radiation is very dangerous and can contaminate an area for a long time. Explosions can happen in many industries and these can kill and injure people and cause suffering and great damage. But, with an explosion involving nuclear material it is not only the blast which causes destruction and harm, the radiation which is released after the explosion continues to be a threat for generations of people. The larger the amount of nuclear material which is present in one location, the greater the damage that an accident will cause. Because the effects of an accident in the nuclear industry are so much more serious than those in other areas of energy generation, the care taken in the nuclear industry has to be very much greater. In fact, when the costs of the development and maintenance of safe nuclear energy generation are taken into account in combination with the very serious health threats, it seems to many that other sources of energy are much more desirable.

If the costs of developing and running the nuclear power industry had been invested in sustainable energy generation, it would have resulted in more energy being produced from a source that is virtually no threat to human health and will never run out. Once established, sustainable electricity generation also has much lower running costs than electricity generated from nuclear energy, because it produces no dangerous and costly wastes. So in terms of cheap electricity generation, nuclear power has fallen far short of early expectations as well as posing a serious threat to very many people.

Perhaps initially nuclear power was seen as a good alternative to oil, and therefore of strategic importance to countries who rely on others for the oil they need, particularly in times of conflict and war. However, with the weaponry which is now available to most countries in the world, nuclear power stations are now a focus for attack. If a nuclear power station were bombed it would not only take away a major amount of electricity-generating power in one fell swoop, it would also kill many people first by the explosion and then by the radiation which would escape from the bomb site. Current sustainable electricity generating methods are much less centred in one location, so attacking one site would not wipe out as much electricity generating power in one go, and more importantly, no deadly radiation can escape from a bombed sustainable energy generating site if, of course, non-nuclear weapons were used in the attack. To stop all research and development work on nuclear power generating now would probably be counterproductive, but this work should only be carried out in installations that contain a tiny fraction of the nuclear material which is currently held in nuclear power stations and nuclear fuel reprocessing plants.

At present work is being carried out on the development of various methods of producing sustainable energy, such as wind turbines, wave energy generators, and solar power panels. Electricity is produced by many of the sustainable energy methods. The main challenge is how to store this electricity efficiently. One method is to pump water to a high elevation using surplus electricity and then run it down through turbines when electricity is needed. Electricity can also be stored in batteries. Another method is to produce hydrogen by splitting water molecules or from other sources using spare electricity after which the hydrogen can be stored. It can then be recombined with oxygen by burning or using hydrogen fuel cells when energy is needed. Storing energy in the form of hydrogen has the advantage that when it is recombined with oxygen, the waste product that is produced is ordinary water. This is in contrast to burning fossil fuels which produces a variety of waste products which can pollute and degrade the land, sea, and atmosphere. Considering that the majority of energy which we use worldwide is derived from fossil fuels, we need to work towards replacing the energy gained from fossil fuels with energy from sustainable sources. In addition to this, many manufacturing industries depend on oil and gas for their raw material, therefore a compelling argument is that the oil and gas we have left should be used to make products, rather than be burned.

Currently, we do use energy from sustainable sources, but the amount we generate in this way is a fraction of what we use from fossil fuels. This needs to be increased before we run out of fossil fuels, particularly if we want to continue to support our current world population and to continue to have the life style that especially the richer part of our world population is now used to. In fact, in order that the poorer section of our human population is made richer and that their quality if life is improved, more energy will be needed than we already use. Besides producing more sustainable energy, we could also prolong our fossil fuel stocks, if we conserve energy more and use it with greater efficiently than we do at present. More efficient use of energy would also mean that our energy needs would be met sooner using sustainable energy. However, unless we radically change our economy, we have got to increase our sustainable energy production to at least equal that produced by fossil fuels to continue our present way of life.

The Natural Religion emphasizes the central role of energy supplies in our world economy. In order that our economy is sustainable in the medium and long term, as well as moving towards eliminating poverty, we will have to replace our current finite fuels with sustainable sources of energy and increase our capacity for energy conservation. **There is plenty of energy around us, what we have to do is develop safe and practical ways to harness it.** Because nuclear fuel is limited, and because nuclear radiation is potentially so lethal and continues to be lethal for such a long time, the stance of The Natural Religion on nuclear power is that research and development in this area should only be carried out in establishments where just a fraction of nuclear material currently held in nuclear plants is present at any one time. When developing and working with nuclear material we have to design our operations on the premise that at some point there will be an escape of radiation. **Nuclear radiation is too dangerous to assume that nothing will ever go wrong, hence the suggestion to keep the amounts of nuclear material in one location very small.**

Our world economy is characterized by cycles in economic growth, followed by economic recession, followed again by growth. Such cycles are accompanied by swings in demand for products, the cost of borrowed money, and changes in the value of money currencies. All of these changes are interconnected and influence the way the economy operates, and therefore also the sharing and distribution of our resources. The ideal situation would be to have a stable economy in which prices, supply of goods, interest on borrowed money, and the value of currencies are at a level at which people can obtain the resources they need. Part of such a stable economy would have to be cycles of change in which different products and services would take over from each other. Change will always happen in any part of the ecology of any species, simply because overall conditions on Earth change all the time. For example, in the case of food supplies, differences in yearly weather will result in variation in harvests from season to season, and as a result our economy will also always go through regular changes. The management and organization of our economy have to take these changes into account, and not allow them to lead to fast growth, followed by recessions.

**The measure of success of an economy is often taken to be strong growth, but growth cannot go on indefinitely because there is a limit to wealth, the products we produce, the amount of money people have, people's needs, and the number of people that the world can support. For these reasons no economic growth can continue for ever.** When people have spent their money, or they simply don't need any more products, then the economy will stop growing. If the economy stays stable or just grows to keep up with an increase in population, then the conditions for poverty-causing economic recessions are less likely to occur. But, the natural gluts and shortages of some of our resources have got to be taken into account when managing our economy. So, a stable economy should have the capacity to change, rather than to react with rapid growth and recessions. Managing our economy in this way is primarily the responsibility of our public authorities.

One aspect of the way our world economy operates is the value of shares in public companies. A shareholding represent a certain percentage ownership of companies, and when a company makes a profit the shareholder receives a share of that profit, or dividend. These shares are bought and sold, and many shareholders attempt to buy shares at a low price and hope to sell them again at a higher price as the economic performance of the company improves. One practice in the trading in shares is to judge the performance of people whose job it is to buy and sell these shares on the profits they make on a three-monthly basis. However, it may take as much as three years for the true economic performance of a company to become apparent. Yet, there can be pressure from the trade in shares to be able to report short-term profits and this can lead to economic shortcuts, or the selling of assets. By taking these short-cuts, or without these assets, a company may actually perform worse econom-ically in the medium and long run and so not contribute to a stable economy. Share prices should be influenced by the medium and long-term economic success of companies, not short-term speculation, and it therefore benefits a stable economy when the profits of share trading are assessed on a longer term than three months.

As part of The Natural Religion's main aim of achieving a good quality of life for our world population, it seeks to help towards creating a stable econ-omy, in which swings from strong economic growth to economic recession are evened out, and in which most people can obtain the resources they need. However, our economy has to change over time, so a stable economy will have to have cycles of change as part of it, but not uncontrolled swings from strong growth to recession and back again.

It is in the interest of the consumer to be able to choose between the products or services from different companies. In this way these companies will compete for business and keep the price reasonably low and quality high. It is important therefore to have more than one company producing the same kind of goods or services. However, if there are too many companies producing the same product, they may be too small or inefficient, or they may produce too much and will then not be able to sell their product. So, once again we have a balance between two extremes. If there are too few companies a monopoly situation exists, and if there are too many they will not be economically viable. Public authorities, once more, play an important role in helping a particular part of the economy to reach a balance which secures economic viability for companies on the one hand, and fair prices and quality of product for the general public on the other hand. This balance may have to shift if general economic circumstances change, but appropriate diversity within economic sectors is essential.

Different-sized companies, as mentioned already, may also be important in keeping diversity in the economy. The relationships between the people operating a family business are completely different from those in a large company that is run by a hierarchy of managers. A commercial operation that is owned by its workforce can have different priorities, compared to one that is not owned by the employees who run it. Certain services may be very important for society in general and may be better delivered by a publicly owned company whose priority it is to deliver that service, rather than by a private company whose priority is to make a profit. Examples of this are health care and public transport. Some products can only be developed and produced by large, often multinational, companies because smaller enterprises may not have the resources and worldwide contacts to be able to produce and market these products. These are just some of the reasons why it is important to have diversity in an economy. All of these different types of commercial operations have their own advantages and disadvantages, they are all efficient in different ways and for different reasons. **In order that the economy as a whole benefits from as many efficiencies as possible, we should create an economy in which a diversity of such companies can exist. A diverse economy also helps to create a more stable economy in the medium and long term when the inevitable economic changes occur.**

What should also be part of the diversity in any economy are organizations whose function it is to carry out research and development. Some research and development is carried out by private commercial operations, but being private and commercial their first priority has to be to generate a profit. Research and development is by its very nature experimental. This means that it may or may not have a positive result, find out something new and/or useful, or is economically viable. For this reason commercial enterprise has to carefully select the type of research and development it carries out as it can only afford to invest time and resources in the type of research that is most likely to have a positive result. This means that the more experimental and explorative research tends not to be carried out by private enterprise, but research like this where the results are uncertain is still very important. Although applied commercial research usually has a greater chance of success, research that is more experimental is more likely to result in entirely new and unexpected findings. Many of our, by now, important discoveries have resulted from unexpected findings of experimental explorative research. For example, in 1896 Henri Becquerel happened to put a package of uranium salts on a photographic plate in a drawer in his laboratory in Paris. After some time he noticed that the photographic plate had been affected by the package similar to the effect that light would have on the plate. He asked his student Marie Curie to investigate this, who then, together with her husband Pierre, discovered radioactivity.

For this reason it is important for our society as a whole that both experimental as well as applied research is carried out. Since commercial enterprise can only afford to carry out applied research, we also need to create and support organizations involved in the more experimental explorative research. The opinion included in The Natural Religion is that the striking of the balance between experimental explorative research and more applied research should be done by public authorities rather than private commercial interests.

It is not only via public authorities that people can influence diversity within an economy. Individuals can also influence the economy directly by the way they invest their savings, and from whom they buy goods and services. **By buying from several commercial operators within an economic sector, people can encourage diversity and competition in an economy.** By buying more expensive and better quality goods or cheaper and lesser quality products we can also influence competition and in this way help to maintain diversity. Individual people can also help to promote an ecologically and economically sustainable economy by investing in funds that only include companies which operate sustainably. The buying power and economic strength of the general public is what keeps the economy going, so we can influence it by buying from the types of companies we want in our economy.

**People can also help to reduce poverty in our world economy by selectively buying products produced by poorer people.** Richer people have more economic power than poorer people, so richer people can make the poorer regional economies richer by buying from them. Certain products from the tropics are marketed as having been produced by independent small producers. These products may be somewhat more expensive than the same product produced by large companies. But, if we buy the products from poorer small producers, we know that we are helping to reduce poverty amongst our fellow human beings, as well as increasing the diversity and competition in these economic sectors which is in our interest.

In considering the diversity within our economic system, we should keep in mind that the main function of our economy is distributing and sharing our resources amongst our world population. While the shareholders own a company, those who do business with it are also affected by the activities of that company and are sometimes called stakeholders of the company. We are, in fact, all stakeholders in our economies. This is another reason why the general public needs to regulate and control the economy to a certain extent via its public authorities. Effectively we are stakeholders in many economic enterprises, and it is in our own interest to create a diverse economy. The more diverse an economy is, the wider choice of companies stakeholders have.

**One of the aims of this book is to help towards establishing a sustainable, adaptable but stable and diverse economy, because these characteristics seem most likely to allow the economy to fulfil its main function of sharing and distributing resources to human beings around the world.** Our economic system is the way we attempt to resolve the basic conflict between the need for all of us to share our world's resources and the need for each of us to have reliable resources so we survive.

**The Natural Religion concludes that the prevalence of severe poverty in the world indicates that the current balance between ownership and sharing is too far towards the extreme of ownership, and that by eliminating poverty and equalizing socioeconomic differences we will improve our economy and create a better quality of life for all of us worldwide.**

As well as influencing our economy by the way we spend our money, we also shape the lives of billions of people by means of politics, which we now look at next.

# Chapter 9

# Our Social Life

> We help, share, and compete with each other
> and billions of people worldwide,
> this necessitates organization

## 9.1: Humankind, a social animal

We humans have developed our social interactions to such a degree that most of us have come to depend on them for our basic survival. Gathering and sharing food, as well as protection and shelter, are everyday needs that involve the cooperation of others. But, apart from these basic necessities, we have also developed an emotional need for the company of other people. To be forcefully separated from all human contact is a very real punishment. The reason why our emotional makeup is so geared towards dealing with other people is that by cooperating we can achieve goals that we cannot accomplish on our own. For example, most of any person's knowledge is directly or indirectly received from other people, and none of the technology that makes our human species so powerful is the result of one person's efforts only.

Social behaviour has been successful in the ecology of many animals besides ourselves. As mentioned in previous chapters, the social insects – ants, termites, bees and wasps – are extremely successful. In terms of their biomass and the energy they consume, social insects are the biggest single animal group in many of the habitats where they live. Individual insects carry out specialized tasks that are coordinated with those of their nest-mates. Together, the social insects are able to achieve what they could not achieve separately and this has allowed them to out-compete many other animals groups. Many species of animals live in groups and benefit from a social way of life, such as flocks of birds, herds of mammals, and shoals of fish.

Social behaviour is based on genetic information that determines what way an animal behaves towards others of its own kind. There are, however, differences in the social behaviour of various groups of animals. The social insects are controlled by their genes in the way they behave. Ants, for example, reacts to specific circumstances in a very fixed and predictable way, but birds or other animals have more choice in the way they behave socially. For instance, as already mentioned in Chapter 5, if one vervet monkey gives an alarm call repeatedly while there is no danger, the others can learn to ignore that individual while still reacting to the same alarm call given by other more honest individuals. We humans have even more choice. We can choose when, with whom, and in what way, to be social in a wide variety of social interactions.

Social animals, including ourselves, achieve more by cooperating and working together, but at the same time also compete for resources with each other. The balance which is struck between caring for themselves and acting socially is largely determined by the genes in most social animals. Our human genes give us more choice in deciding where the balance lies between our social and selfish behaviour than is the case with other species. Theoretically, each person could make a slightly different choice, and if this were to happen, the contributions that people make towards running their society could be uncoordinated and disorganized. But, by combining people's efforts, we have developed ways of organizing entire societies so that their social structures function in an organized and efficient way. An example of organized contribution to social organization is road use by busy traffic. Each of us contribute towards social organization by only using one half of the road so that people going in the opposite direction can use the other half. This is much more efficient than if all of us used both halves of the road.

Social organization has been part of our human ecology for such a long time that we take it for granted, but it has changed as our ecology changed. The ecology of hunter-gatherers was characterized by very low population densities occupying large areas that provided them with food. As humans started to practise agriculture, we had much higher population densities and smaller areas in which land tenure and ownership allowed food supplies to be predictable. However, agricultural food stores did become a target for outsiders to attack and steal. So, agricultural society as a whole had to organize itself to share out, protect, and secure the land and water that produced its food. Ever-increasing populations and specialization of work brought further social coordination and efficiency. The high level of social organization that we have now therefore developed since the advent of agriculture. **This means that our current socially complex way of life, while ultimately still based on our genetic makeup, is more directly based on knowledge. Our social behaviour has changed much more in the last 10,000 to 12,000 years than our genetic makeup changed in the same period, hence the development of our social behaviour since the start of agriculture is often called our social evolution as distinct from our genetic evolution.** We are not genetically programmed to live together in high density and highly socially integrated communities, we have learned to do this and now accept our complex social organization as a normal part of our lives. This high level of social cooperation generated by our knowledge has given us a powerful position in the system of life on planet Earth.

Therefore, The Natural Religion sees our social behaviour as part of our human nature, and concludes that our current complex social systems are relatively recent and so far successful developments in the history of our species.

## 9.2: Leadership

Agreement on what we are going to do, and on how and when we are going to do it, makes our social behaviour efficient. To make the process of reaching agreement more efficient, we have leaders. One leader, or a combined leadership of several people, assess a situation and make a decision that the rest of the group follows. Leadership is also found in the social behaviour of other animals with one or a small number of usually socially high-ranking individuals making decisions about the daily choices that most animals have to make. Baboons, elephants, chimpanzees, and many monkeys species all have leaders who decide about such daily decisions, as which area to go to for food and where to sleep at night. In our own case, leadership is, of course, more complex.

We have to share the Earth's resources, and part of the function of social leadership is to make decisions about sharing such resources as access to fertile land and usable water, and also what sort of ownership of resources we have. The leaders of a community can have great power and influence over many people, and their decisions have far-reaching consequences for people's welfare and quality of life. A fair sharing of resources may not necessarily mean that everyone gets exactly the same amount, so leaders often have to take decisions that some people may not like. Sharing resources can mean limiting the amount that individuals can receive, in order that everyone receives enough to survive. Decisions that involve how much people contribute to the common good, as in the case of taxes, are often not popular. But, the resources that are needed to run and support a society have to be gathered in some way. Most people want to benefit fully from the services that their society provides for them, but only want to pay the bare minimum in taxes. One of the most important tasks of human social leadership is to treat everyone equally and fairly, but if leaders of a society treat some people unfairly compared to others this can result in violent reactions threatening the security of that society.

Amongst hunter-gatherers and early agricultural communities, leadership was less important than it is now. Currently amongst groups that live those kinds of traditional lifestyles, leaders may be no more than figureheads with limited powers over decision making. However, as agriculture developed, populations increased and society became more complex. As agriculture advanced it became increasingly possible to accumulate wealth, and as leaders of these societies had more power they were also in a better position than others to become wealthy and many did. The need to fight and defend a territory and its resources against others also became important as agricultural wealth accumulated. Wealthy and powerful leaders controlled not only the economy but also bands of armed men. So, their wealth allowed them to impose their will by force on the general population. Thus, a type of leadership developed that did not depend for its power on agreement with the community, but on the wealth of its leaders. This meant that they were not accountable to the general population. These rich groups of leaders also stopped the poor from growing wealthier. By keeping themselves wealthier than most other people, the leaders retained power in their own hands, leading to the creation of different socioeconomic classes.

Besides using economic measures and physical force to maintain social power, leaders also used religion to enhance and confirm their social status. Some claimed special knowledge and understanding of a religion that other social classes did not have, or put themselves in a position of special confidants and communicators with a god or gods. Others accorded themselves the status of supernatural beings or gods. All of this reinforced and widened the social difference between the leadership class and the common people, and protected the leaders' position of social power and wealth.

The social distance between the wealthy and the poor became so large that people were even owned by other people in many societies all around the world. During a period in the development of our social evolution, economies all over the world depended on work that was carried out by slaves. To this day, societies remain in which some people are effectively slaves for economic and other reasons. These types of social systems are very far removed from our original egalitarian hunter-gatherer societies.

Despite the wealth and power of autocratic leaders, people continued to desire egalitarian societies. Types of leadership developed that involved the general population in decision taking, these democratic systems developed in a number of locations around the world. Usually a group of people were chosen or elected to be leaders by part, or all, of the population. These leaders were assumed to make decisions to the liking of those who elected them, and were only given the power of leadership for a limited period of time. Another feature of many democratic systems is that leaders while in power stay accountable to their electorate. In other words, if they make too many unpopular decisions, or if they make too many mistakes, they can be instructed to change their policies or to give up leadership altogether.

As briefly mentioned in Chapter 2 and as with so many other aspects of our lives, the various democratic systems that we have developed during our social evolution, have tried to achieve a balance between two unstable extremes. One extreme is a powerful leadership that is only accountable to itself. Strong unquestioned leadership can give great efficiency to a social system and can bind a population together so they act as one strong and united group. Such a nation can be very successful in its own defence as well as with its economic activities, but a strong leadership that is not challenged usually turns into a dictatorial leadership. One problem with this is that when a dictator makes a mistake (and no person ever makes no mistakes), it can be very serious because these decisions affect so many people. Another problem with the mistakes of a dictator is that they often go unchallenged, as autocratic leaders tend not to tolerate criticism. Yet another result of dictators maintaining their own wealth and power and large differences in wealth within their societies, is that poorer people tend to feel helpless and frustrated as they have few choices and little say in planning their own future. The success or otherwise of a society depends on what its people can accomplish, and if a dictator controls a society, quite often people's individual expression and initiative is suppressed. This means that a significant number of people cannot reach the full potential of their capabilities and this retards a whole society. So, there are a number of factors that make dictatorial leadership unstable. In fact, dictators can usually only maintain their power over a society with the threat of violence, which, of course, seriously compromises people's quality of life.

The other extreme is that every member of a society takes part in every discussion about every decision that has to be taken in a society. While this is very egalitarian, it is not practical or efficient with the size of our current communities. If people disagree, or have alternative solutions, good and timely decisions would be very difficult to reach. This would even be the case in communities consisting of hundreds rather than thousands of people, in which it could theoretically be possible to have mass meetings on a regular basis to discuss decisions. To involve every person in every decision in a community of thousands of people or more is simply not possible. In a social system without any leadership most communal activities are very inefficient, as well as being vulnerable to violent attack by more organized groups of people.

Democratically run communities try to combine the advantages and avoid the disadvantages of both extremes by having leadership composed of a limited number of people who have power for a limited period of time only. Most decisions are group decisions, which means that mistakes or bad decisions will be noticed by people who are in a position to challenge them. On the other hand, it also means that any discussion about a decision is carried out by a small enough number of people so everyone has an opportunity to hear the opinions of others. This means that decisions can be discussed and considered beforehand and then be taken on time.

The fact that leadership is only for a limited period of time means that a greater variety of opinions and approaches influence the organization of a society when new leaders take over. It also limits the time period in which bad leaders can make bad decisions and corrupt leaders use their power to accumulate wealth. A disadvantage is that good leaders are also replaced, but by replacing leaders regularly the general population can keep some control over them. This is one feedback system of control over the leadership.

One very important element that is needed in democratic social management, is that the general population is reliably informed. If the public does not know what the leadership is doing and why, they cannot make informed judgements on their present leaders or on electing future leaders. In smaller communities, word of mouth can give people the information they need to judge their leaders. However, in bigger communities the media (such as newspapers, radio, and television) gather and give out this information to the general public. Therefore, it is very important for the proper operation of a democratic system that the media are free to gather correct information, and that they reliably inform people about it. In societies with dictators as leaders, the dictatorship usually tightly controls the media; this is because it usually wants to hide errors and unfair and corrupt management from the general public.

Another aspect of democracies is that opinions are allowed to be freely expressed that are contrary to those of the leadership. Any member of the general public should be able to say that they disagree with the leadership without being penalized in any way for this. In fact, in most larger democracies a group of people, the parliamentary opposition, scrutinizes and criticizes the leadership and are part of the social management structure of society. If a leadership makes mistakes it is the function of the opposition to draw people's attention to this. The opposition can also help the leadership by joining in constructive discussions and help them to reach decisions. A democratic opposition should itself also be open to the general public for their judgement, being part of the democratic system.

The amount of power we allocate to the leaders in a democracy is, like so many other aspects of our human ecology, a balance between two extremes. This balance exists between, on the one hand, the power of democratic leadership and, on the other hand, the feedback system of control of people over the leadership. One way this balance can shift is in the amount of people it takes to elect a representative to the leadership. The more people are needed to elect a representative, the less individual people's opinions can be taken into account, and the more only a generalized view of the opinions of the general public is reflected in decision making.

Another factor that affects the democratic balance is that democratic systems don't allow the entire population to elect leaders. All during the development of democracy, and up to the present day, only people of a certain age are allowed to vote, up to very recent times women were not allowed to vote in many democracies, and in many earlier democratic societies only people with a certain amount of wealth could vote. Limitations to democracy can also exist in the type of people who are allowed to be elected. In some democracies people may only be an elected representative if they are judged to be of good character, and what is judged to be good character usually depends on the particular culture and historical era. A trend which still exists in some democracies is that the process whereby a person becomes a candidate for the more powerful leadership positions may need so much resources that only the wealthy can afford to become candidates. This type of democracy tends towards the extreme with less involvement of poorer people in social management.

Although different democracies strike different balances, on the whole, democracies do allow people more freedom to attain their full personal potentials and gives them more choices than dictatorial political systems do. **The opportunity to realize our full potential means that we are more likely to achieve a good quality of life, and that more people will be economically successful. At present if we compare all the countries in the world; the wealthier the country, the more likely that it is a democracy.**

The balance that most democracies try to attain is to have leaders who are efficient, backed up by a system that helps them to make good decisions, but who stay accountable to the population whose social system they manage. **This balance tries to combine the equality and general involvement in social management of most members of a community, which the hunter-gatherer social system also had, with the efficiency that leadership brings to large populations.** During the time that we evolved as a species our social system was in all probability akin to the hunter-gatherer system. Therefore, we are most likely genetically best suited for that type of social management, hence our urge to try to get as close to this type of social organization as circumstances will allow. **Democracy is therefore a knowledge-based adaptation of our genetically determined social system.**

**Amongst the conclusions which are part of The Natural Religion is that the development of democratic social management has as its basic aim a fair distribution of the Earth's resources.** Three functions have developed in democratic systems to try to achieve this. The first, is an efficient and competent leadership that makes good and timely decisions. The second, is a general public that has influence on decision making and also has sufficient control over its leadership to make them accountable to the society they lead. And the third, is freedom of the media to allow them to gather and disseminate correct information, so the general public can make informed judgements about their leadership. Linked to this is the right for people to express disapproval of their leadership without being penalized for this in any way.

**The ideas discussed in this section leads The Natural Religion to recommend a democratic system with a large measure of control over leadership and social management by all adults, men and women, rich and poor, while still allowing leaders to organize society efficiently so that people have the freedom to reach their full potential.**

## 9.3: Political compromise

As we know, we need resources and we need to divide them fairly between us, and one of the main functions of leaders is to organize the sharing and distribution of resources. Because our resources are limited, we cannot get everything we want, so our social leadership has to compromise and attempt to share our resources fairly between us. Compromising means that we get some of what we want but not everything. It is easier for a dictatorial leader to deny someone something than for an elected leader. In a democratic system leaders have to try and reach agreement amongst those who have to compromise and this can be difficult. For democratic leaders to be able to lead effectively they have to attempt to reach agreement amongst as many people as possible and this has given rise to leaders forming groups amongst themselves. Such groups or political parties reach a consensus about policies and decisions that makes it easier to lead a society. The process whereby compromise agreement is reached within and between parties we call our political system.

Political alliances are not only a feature of our human social system, but also occur in other animal species. Individual chimpanzees, gorillas, baboons, vervet monkeys, and lions all form alliances or friendships, and support each other in acting as leaders. Political alliances amongst chimpanzees can be very complex. The physically strongest male is not always the top leader. There are many examples observed of a male chimpanzee becoming a top leader because he was better able to form political alliances with other chimpanzees, than another physically stronger male. The top leader on his own may not be able to dominate the strongest male but with the support of his allies he can. The chimpanzees who support a leader do not themselves become top leaders, but do rise in status because they are supporters of the leader. In fact, there is an example of one older male chimpanzee who lost his leadership to two younger and stronger males, but who continued to play off the two younger males against each other by supporting each of them in turn. In this way the older ex-leader continued to be the power broker in the group. In a further development, the older ex-leader and the weaker of the two younger males eventually attacked and killed the stronger of the two young males. The weaker young chimpanzee now became the permanent leader with the ex-leader as second-in-command. It is interesting that the older ex-leader eventually sided with the weaker of the younger males against the stronger, as he had more control and power in the group as a whole when the weaker younger male was leader than if the stronger had been leader. Such strategic shifts in political alliances amongst chimpanzees are very like our own political relationships. This is another aspect of our social system that appears to have its roots in our evolutionary past.

As we have seen, agriculture gave rise to bigger populations that also resulted in the need for leadership and politics to reach compromises about sharing resources. Many of the early political systems were no larger than an ancient city plus the surrounding area where farmers produced food. These early city states tended to be more dictatorial than democratic and the leaders generally made themselves wealthier and had more political power than the general population. As our social evolution developed, city states fought each other over access to resources such as clean water, fertile land, and food stores. Then cities formed political alliances with each other, so together they could attack or defend against other larger political alliances. Since then, political units and their human populations have considerably increased in size, currently nation states can include many millions of people and cover most of a continent. In such large countries the political system is big and complex. People can feel removed from their political leaders if there is a large population spread out over great distances, but big political systems usually have subdivisions within a state and in this way social management is brought back to a more local level.

Political systems have an important influence on the way resources are distributed. In strict dictatorships, virtually the entire population depends on the whims of the leadership for their livelihood. However, during our social evolution there has been a tendency away from dictatorships and towards democratic political systems. Even though democratic systems work in apparently complicated and ponderous ways, they do result in a fairer distribution of resources and more freedom of choice for people. At times, one country with a democratic system may have a political alliance with another country which is not democratic. The function of such an international political alliance is the exchange of resources in the economic interest of both countries. Political alliances exist to reach agreements and compromises, and have to try to reconcile different interests to reach a solution which is accepted by all concerned.

Politics is often criticized or berated for being insincere and two-faced. However, when two sides of a dispute cannot reach agreement, the job of politicians is to keep the negotiations going until an agreement is reached. The alternative is that a dispute could develop into a very entrenched situation in which the opponents don't talk to each other, or that it escalates into violent conflict. To keep two sides to a dispute talking may require great diplomacy on the part of political leaders that at times may appear to be two-faced. In a democracy, political leaders, on the one hand, have to make some unpopular decisions, while, on the other hand, they must persuade the general population to elect them to power again. Added to this can be the criticism of opposition parties, who want to get into power themselves and claim that the governing party is not sincere. The root of a lot of these criticisms is the necessity for the political system to be able to reach compromises, and not being able to give everybody everything they want.

To convince people with differing interests, leaders both in democratic as well as some non-democratic social systems have to reach compromises about the crucial matter of sharing and distribution of resources. To accomplish this, agreement has to be reached and the system of political alliances and parties is the way in which people support each other in order to reach these compromise agreements. **The Natural Religion urges politicians to reach agreement by negotiation and to use diplomacy to achieve compromises, rather than to allow crude conflict to alienate opposing sides.**

## 9.4: War

To return to basics, people compete for resources and it may appear that the easiest way to get more is by stealing from others. **Since the lack of resources does threaten life, people have to acquire and safeguard**

**them, and for this reason people will resort to violence to obtain or protect essential resources for life.** As always, we have to strike a balance between our own interests and those of others. Some people try to move this balance too much towards themselves by trying to get resources by stealing or acquiring them unfairly some other way. To act unfairly like this is, of course, a mistake and will often involve breaking the rules of a society. Not only have we developed police forces and armies to stop people from breaking the law within a society, but armies also stop people from outside stealing our resources as well as attacking and stealing resources from other communities. Large armies of people fighting, stealing resources, and killing each other has been common ever since the dawn of agriculture. While the stone-age Cro-Magnon people did make drawings of people fighting and killing each other on the walls and roofs of caves, no defensive works of any description have been reported around early human settlements, but defensive structures are found near old settlements of early agriculturalists. Whether or not our hunter-gatherer ancestors fought with each other about the size and location of their territory, they most likely did not have accumulations of wealth and resources such as food stores as their whole territory in which they hunted animals and gathered roots, fruits, and seeds was their food store. Agriculture's food stores do form a very tempting target for others to steal, and from early on armies have often fought each other over them.

Since the advent of agriculture, we humans have been at war with each other somewhere in the world more or less all the time. Why do we do this? We are a very social species, yet we spend a lot of time, resources, and effort in trying to militarily dominate and kill each other. We used to live in groups of less than one hundred people, we now live in groups ranging from many thousands to millions of people. When such a change occurs in the population density of other animal species it normally results in severe overpopulation. One phenomenon associated with overpopulation in most animal species is an increase in aggression amongst the members of that species. In the case of our own species, we have adapted to our increased population densities by developing ways of coordinating many specialized tasks that enabled us to live closely together, and because of this coordinated specialization we have become very dependent on each other. What happened, however, was that this specialization was also used to wage war by forming coordinated armies that were trained to fight. As our societies grew bigger, so did our armies and so did our wars and the numbers of people who were killed in them. **Our social evolution is still at the stage where separate social units such as ethnic groups, nations, and military alliances of many nations attack and kill each other on a regular basis.**

Not everyone in the world has the same amount of resources, in fact, millions of people die due to food shortages. It is normal in the ecology of many species that they will attack each other when resources run short. Ants, for example, have extensive wars with each other. Ecological imperatives, such as food, apply to us humans as much as to any other animal species. We do compete for many resources. However, it is also part of the behaviour and ecology of other animals to avoid the harm and damage caused by physical violence. Posturing and threatening with visual, sound, and odour signals are used in conflicts, but violence causing actual physical damage is often avoided in the case of many species of animals. While we humans also try to avoid physical conflict if at all possible, we also wage a lot of war. One rationale for our warlike ways is, that our ancestors survived because they were prepared to got to war and were successful. People with warlike genes killed those without these genes so few of these survived, leading some to conclude that we inherited violent genes which we can't do anything about and that we are an innately violent species.

**A study of European countries examined between 275 and 1025 years of their history and found that they were at war an average of 50% of the time. Another study concluded that in the last 5600 years there were only 292 years, during which there was no war anywhere in the world. Between 1820 and 1945, approximately 59 million people were killed in wars. In the 20th century there were between 130–140 million war-related deaths, with 50 million killed and more than 200 armed conflicts taking place from the Second World War to the end of the century, most of them in developing countries.**

**So yes, we humans are very warlike.**

Why do people wage war? There is a saying 'War is politics carried out by other means'. Is fighting a part of the political process? In the history of most countries of the world this has been the case. Amongst our smaller traditional societies, which we assume are similar to those of our evolutionary ancestors, some are quite warlike but others are not at all violent. Some traditional cultures abhor violence of any type, and have many ways to resolve conflicts in peaceful ways. This shows that even though most people are capable of violence, we also have the capability of creating cultures and social management that do not include violence. So, we have proved that we have the potential to live in peace with each other, as have other species of animals.

The reasons for wars are often either directly about access and control over resources, or about secular and religious political power. Conflicts over resources such as clean water, fertile land, oil, and trade routes have many times resulted in war. Wars about political power are usually also ultimately about control over resources. Political power is often about controlling what people think, and this can be very successful. One demonstration of how successful this can be, is the frequency with which political leaders have been able to persuade people to go into battle to kill other people and at the same time put themselves at risk of being killed. Such control of a society gives leaders power. War and controlling and dominating other people gives great political power and has been part of our political systems for thousands of years.

**War involves maiming and killing people causing a huge amount of suffering, anguish, and grief. It destroys families and relationships, it cuts people's lives short and shatters the lives of many of the survivors. War has reduced and blocked the potential of many millions of people. War destroys life and quality of life.** War also negates our achievements; it destroys buildings, bridges, and roads. It also destroys agriculture by burning crops, wrecking irrigation systems, and simply by stopping people from carrying out farming. Soldiers also steal food and burn food stores. Knowledge has also been lost due to the turmoil of war, such as by killing the people who possess learning or by destroying libraries full of books. Much of the knowledge of a particular culture can be lost if that culture is crushed by war. Our wars are cruel, and needlessly wasteful of human life and of our achievements.

Because winning is so important to both sides in a war, people go to great lengths in order to triumph. This has meant that many inventions and much ingenuity has been achieved as a result of war. Some of these achievements that were initially used for military purposes, have also proved useful for peaceful activities. Methods of architecture and general construction, and many technological inventions, such as radar, are examples. However, our knowledge has taken us so far that we have also produced weapons that can kill millions of people. Some of these weapons can not only kill whole populations, but also poison the land that we rely on for food. Weapons such as bacteria and viruses causing lethal diseases have the potential to kill millions. Nuclear weapons not only kill directly when they explode, but also continue to kill for many years afterwards with the dangerous radiation that they release. Our ingenuity has produced nuclear bombs that are so big that they could change Earth's weather patterns on which agriculture and our food supplies depend. **Is our knowledge-based human evolutionary experiment destined to fail? Has our very powerful knowledge now gone so far that we will destroy ourselves with it, our own ingenuity bringing about its own downfall?**

Despite our tendency for fighting and to wage war, we also strive towards peace. During wartime, people cannot lead emotionally nor physically sustainable lives. Many resources are needed for war and these resources run out, so most wars can only last for a number of years. They then stop, so people can recuperate and replenish food supplies and other resources.

After some time of war the desire for peace increases. What the warring factions hope for is that after the war they will be in a better position to come to an agreement than they were before the war. However, if a compromise agreement had been reached before the war, both parties would have been better off. In wars, many people's lives are wasted and many resources are squandered. A nuclear war is the ultimate demonstration that the results of war are infinitely worse than the problem it was designed to solve. All this waste would not have happened if a compromise agreement had been reached before war was embarked upon, and both parties would have been much better resourced to proceed with further developments. Unless a compromise would somehow also result in the loss of many human lives, the concessions that have to be made as result of a compromise are almost always far less than what is lost and destroyed as a result of war.

To reach a compromise, both parties need to come together and reach an agreement. If one side does not want to reach an agreement and attacks the other side, the side that is attacked has few options except to retaliate and defend itself or let the other side take over. Such takeovers can also result in further wars if the defeated side gains political allies or resources with which it can revolt against its oppressors, thus setting up cycles of violence that repeat themselves.

One way or another, most warring factions do eventually want to establish peace. Tragically, such a peace is usually prompted by the horrors of death, grief, and suffering. If this wish for peace had been felt before the war or series of wars, none of the waste of life and destruction need have taken place. War seems to be our way of giving ourselves strong motivation to have peace, and determination to reach compromise agreements. It seems to have been our way of teaching ourselves how much worse the alternative to a compromise agreement really is, and apparently we have to teach ourselves this lesson over and over again!

There are many people and international organizations who help with trying to avoid wars. Where economics are at stake, peacemakers will find it easy to gain many political allies. Even after a war, compromises normally have to be reached anyway, particularly if long-term stability is to be achieved. However, the resources that are spent on war have made wars a whole economic sector in itself. As mentioned in Chapter 8, the weapons industry is thought to be amongst the first three or four biggest industry in the world. It is difficult to exactly pinpoint the extent of the weapons industry, as much of it is kept secret. With such economic powers vested in making weapons it is perhaps more understandable why we continue to have so much war. People have always been very hesitant to compromise their economic advantages and are very inclined to engage in war if they think it would be of advantage to them. Despite the destruction and horror of war, leaders can be inclined to engage in violent conflict as a 'quick fix', rather than try to reach a compromise by means of protracted negotiations. This is particularly the case if they think they are stronger than the opposition, and can win. The full cost of war is only realized after the war is over. Leaders tend not to fully and openly compare the total costs of a war, with the cost of a negotiated compromise before a war starts.

**The Natural Religion is wholly and completely against war and sees it as a cruel, outrageously wasteful, and extremely primitive way of resolving basic competition and conflict over resources.** One of the aims of The Natural Religion is to help to reintroduce and support sophisticated and peaceful ways of reaching compromise conflict agreements, that are capable of dealing with our high human population densities. A system of peaceful conflict resolution needs to take place before violence gets a hold, because if one is physically attacked, more often than not the reaction is retaliatory violence. The threat of the ultimate sanction of physical force by police forces and armies is part of the safeguarding of social structures in most of our societies, but the hope included in The Natural Religion is that we will seek to reduce the use and threat of physical force both internationally and within our communities. International conflicts can be avoided by mediation. **The full costs of war both in terms of human suffering and destruction of resources, compared to the possible costs of concessions made as part of a compromise agreement, need to be clearly identified and recognized in any potential war situation.** In order that we reduce our dependence on police forces and armies to uphold the laws of societies, we first have to change the way social control operates, away from one ultimately based on physical force, towards one based on a fully informed understanding of the long-term benefits of peaceful compromise agreements.

**All through this book it is stressed that distribution and sharing of resources is very important in the avoidance of violent conflicts and wars. Therefore, economic strategies that reduce poverty and guarantee sustainable sharing of essential resources should be part of international human ecological management.** As we have seen, differences between rich and poor, and unequal distribution of resources have resulted in wars many times in our history. Unfair economic treatment of particular groups usually results in social stresses and conflicts within communities. The approach of The Natural Religion is, therefore, to address these root-causes of crime, social conflict, and outright war by correcting imbalances in the way we share our resources. This includes that the economic sectors with an interest in war need to be part of a realignment and enlargement of our economy, that a peaceful reduction of poverty has the potential to bring about, rather than being threatened with economic ruin. **The arms industry needs to be given a vested interest in peace. The effort to reach a more sophisticated manner of national and international conflict resolution will need worldwide political cooperation, and needs to take place before self-perpetuating cycles of violence and counter violence can start.** It is The Natural Religion's hope that public management and leadership, private economic interests, and the general public will all realize that it serves all our long-term interests if this global cooperation is reached. Such worldwide peaceful conflict resolution will be greatly helped by sustainable use of resources, against a background of human world population management and control. These basic ecological factors are important in the control of aggression and violent competition between the members of most species of animals, but **other animal species living in high population densities have developed ways to avoid physical violence amongst themselves, therefore so can we.**

## 9.5: Future politics

Our politics ultimately spring from the fact that we are social animals, and it has had a major influence on our present ecology and the way we live. We did not need our current level of social organization when we lived in smaller groups, but now we live in such high densities that we depend on our social organization and management for our survival. We have already seen examples such as ants, lions, apes, and monkeys that also depend for their survival on social organization and politics. By trial and error, and the use of knowledge, we reinvented many solutions for living in high densities that had already been genetically evolved by other animals. So, politics has a long history in animal social behaviour, and our basic politics would therefore seem to be inherited rather than reinvented. Our current very powerful human ecology is deeply influenced by our politics and the agreements and decisions that it produces. The way politics operates is crucial to the future of our ecology.

As our social evolution and management depend on knowledge, we have much more choice about our politics than animals whose political behaviour is more directly genetically controlled. People have chosen and tried different types of political systems and leadership at different times. These choices have various advantages and disadvantages, and this is the way social evolution progresses. Like our genetic evolution, our social evolution is an experiment the results of which are not certain yet.

However, one important difference between genetic evolution and social evolution is the time factor. Social evolution takes place at a much faster rate than genetic evolution because it is knowledge based. Our knowledge-driven social evolution has been a boom success story over just 10,000 years, which is a very short time compared to genetic evolutionary development. Social evolution carried us very far very quickly, as shown by the growth in the world's human population size in that time, and the influence we have so far had on our environment.

As pointed out before, if there are problems with our knowledge-influenced ecology, we also need to solve them using our knowledge. **We have to assume that to rely on genetic evolution will be much too slow to solve problems created by the influence of knowledge on our lives.** Our politics and the political decisions we make, will have a great influence on how our species will survive in the future.

Our social management has become so organized and complex that people often feel that their opinions just don't count. In our densely populated societies one person's opinion may often not even be noticed by others, let alone have any impact on the way society is managed. If we wish to influence the future management of our human affairs we need to communicate our ideas to other people. This can be done in various ways via the media, including forming groups who support each other in suggesting ideas to society in general.

In dictatorships, however, suggestions on how society should be run may not be allowed by the leadership, and therefore new ideas and questioning of old ones by the general public are ignored and therefore wasted. This waste of ideas is one of the weaknesses of dictatorial political systems. In societies where the expression of new ideas is allowed, the general public can influence their own future society, but because our populations are big we need to participate in the political system and/or use the media to get the message across to enough people so that social management is influenced.

In order to be able to judge the way a society is being managed, and make informed suggestions as to how it could be improved, people need to have information. People have increasingly gained knowledge, and so our understanding of the effects of various political decisions has also increased. As we became better educated and accumulated more reliable knowledge, the tendency towards democracy in social management increased. In the past, only rich people could afford an education and therefore leaders tended to come from the richer socioeconomic classes. The majority of the public were poor and undereducated and were therefore at a disadvantage to either assess their social leadership or become part of it. In those days, political systems tended to be less democratic. As our general level of education increased, the majority of people were in a better position to judge how well leaders cared for society. This inevitably led to a wish for greater involvement in social management by a growing number of people. **So, it would appear, therefore, that future political systems which reduce and eliminate poverty, will also have a better educated population and an increased quality of life, and will most likely be democratic in nature.** As we already saw, democratic systems do have a certain level of inherent inefficiency, and are based on a balance of the various interests in society. The interests of the general public, the public authorities, and private economic enterprise need to be balanced; the self-interest of individual people and the interest of society in general need to be balanced. In democratic systems the people should ultimately decide the type of balance that suits their whole society best. So, for the general public to be able to make choices they need to be educated, and be made aware of all information to allow them to make informed judgements about the future of our politics.

As we saw in previous chapters, we need not choose to allow violence and aggression to shape our future – we need not use war as politics carried out by other means. We can learn from peaceful ways of resolving competition for resources from some of our own traditional cultures as well as from the ecology and behaviour of other animals. **We need to resolve conflicts and reach our compromises about our resources before violence and aggression forces us to wage war on each other. We need to identify potential future wars and conflicts, and solve them before people decide that the only way out is physical violence. We need to use more sophisticated ways of dealing with competition in our ecology than crudely using war.**

Competition for resources is one of the main reasons for conflict. In order that in the future there are enough resources for everybody, we need to make sure that we make sustainable use of our resources. We also need to make sure that our population does not increase so much that we outgrow our resources. All living species have to match their population to their resources, this is a biological imperative. So, to avoid situations that could lead to war, we need enough resources to go round, and we need to make sure that our resources are distributed fairly to those who need them, and reduce and eliminate poverty because poverty without hope causes war. These are aims that we can bring about, but people from all nationalities and from all over the world will have to cooperate in order to have a peaceful future.

Historically the land areas included under central political control have waxed and waned, but generally speaking the trend has been towards fewer and larger political units rather than towards more and smaller units. In recent times separate states have also come together forming larger political coalitions, economic unions, or federations.

As we have already seen, one aspect of larger political units is that the individual person who lives in a large nation can feel further and further removed from the leadership of his or her society, and can feel that they don't have any real influence on what happens in their country, even if they vote for certain leaders in a democracy. The considerable proportion of people who choose not to vote in many democracies is an indication of this. In some elections less than half of the electorate may exercise their right to vote.

However, there are many problems that people the world over have to deal with together. Sustainable use of resources, global pollution, world economics, control of our human population, elimination of poverty are all international problems and need to be addressed internationally. Much international effort will be necessary in our future politics if we are to reduce and eliminate wars. If people around the world see themselves as part of a global community and feel they are involved in international affairs, it may facilitate reaching compromises and so avoid wars. Our future politics will need to be able to address our human affairs internationally, but it will also need to maintain contact with the many millions of people who form each community. This is another balance which our social management needs to achieve.

In summary, our politics past, present, and future is a normal part of our human social organization. We rely on each other's cooperation and help to form our future politics, just as we have in the past. Emphasis of the role which politics will play in shaping our future lives and ecology is part of the ideas which form The Natural Religion.

**Amongst the conclusions arrived at here is that democratic systems have the best potential for striking a balance between the interests of the individual and those of the common good, as well as between the efficiency of centralized leadership and control of the leaders by the members of a society.**

Although conflict and competition over resources is a normal part of our ecology as it is of other living species, many examples exist in our own cultures and in the ecology of other animals for resolving conflicts without physical violence. National and international future politics should aim to identify and peacefully resolve conflicts before a cycle of violence and counter-violence can gain momentum.

One of the aims of The Natural Religion is to help towards increasing the participation of the general public in politics and social management, hand in hand with improving education and general quality of life. As we saw, the richer countries in the world are also predominantly democratic. **People need to make sure that they gain correct knowledge and make informed judgements, so they can take an active role in our political processes and influence changes in our social management.** If enough ordinary people support each other's ideas, powerful interests such as strong leadership and influential commercial interests need to pay attention, and the political system should be such that it allows this to happen. Both political leaders and private enterprise, no matter how big and powerful, rely on the general public for their existence.

As explained in previous chapters, one of the set of ideas which make up The Natural Religion is that essentially the meaning, significance, and importance of a person's life is to support and help care for the future of humankind. After all, we who are alive today are benefiting from the achievements of our predecessors. **The relevance of the lives of those who have gone before us is in our own existence. So also is safeguarding the future welfare of humankind, the ultimate meaning of our own lives.** Part of this is making informed judgements about our future and constructively influencing our future politics.

While politics is about managing communities, religions also focus on each person's individual experience of life, to which we return in the final chapter.

# Chapter 10

# Our Consciousness

🐦 Our brain creates our reality 🐦

## 10.1: Our mental needs

Of all living species on Earth we humans have the most developed brain, but, one way or another, all living beings react to their environment. Plants grow upwards towards the light while their roots grow downwards into the ground. The tiniest microscopic animal reacts to temperature and humidity and can distinguish between food and other objects in its environment. Certain other animals can hear, see, smell, feel, and taste better than we can, but our human brain is capable of greater understanding than that of any other animal species. The knowledge that we can store in our brain, and our understanding of its significance, is unequalled in Earthly life. Because of our brain, we have a consciousness of ourselves and the world around us. As described in the Appendix, our consciousness consists of bioelectrical and biochemical reactions in our brain. Life has evolved bio-electrochemical processes that model itself and its surroundings, this is in essence what life's experience is. Despite the various supernatural explanations that traditional religions suggest, no verifiable evidence exists which indicates that anything else besides chemical processes produce our consciousness and the way we experience life. The chemical reactions that make up life have become so organized and complex that they have reached the amazing achievement of being able to feel, experience, and know about themselves. This complexity, which includes having knowledge and being self-aware, is truly awe-inspiring. Each person's consciousness is what we feel our personal identity to be, our consciousness is us.

Besides our basic needs for survival, we humans also have mental needs. Our brain, like other organs, needs the right type of stimulation to develop properly so it can function to its full potential, therefore the stimulation that causes our brain to register happiness, contentment, pleasure, and joy are a very important part of our consciousness.

We need our brain for our everyday life, and because it is designed to deal with the considerations, the thinking, the problems, the emotions, and the decisions of our lives, it also expects and needs this type of stimulation. Like any part of our body, our brain needs to be used in order to function correctly. While our brain needs to be stimulated, it should not be overwhelmed with sensory input. If we don't have enough mental stimulation we can feel bored, frustrated, or discontented. Forced sensory depravation, such as happens when prisoners are held in solitary confinement, is very stressful and can cause serious mental anguish. On the other hand, if we bombard our brain with too much sensory input, we will not be able to process all of it and we will feel stressed, confused, and not in control. Another mental need is to feel loved, which we get from the company of friends and family. We also need to feel satisfied with what we accomplish in life and feel successful in what we try to achieve. We also have a mental need for enjoying aesthetic pleasures such as the beauty of nature as well as music, literature, and other forms of art. The state our brain is in, is the way we feel. Of all the organs in our body, we are most aware of our brain. Therefore, after our basic needs of breathing, drinking, and eating, our mental needs are felt most acutely. Our mental needs are crucial to our lives.

Like other religions, The Natural Religion deals with our innermost sense of ourselves, which we produce with our brain. **Our brain's consciousness is the core of every person's sense of identity and self-awareness, our mental needs are crucial to our normal and healthy functioning as human beings.** The Natural Religion addresses the needs of our whole consciousness; needs made up of the interplay of our emotions, our ideals, and our ever-changing store of knowledge. Having human ecology as foundation for much of its reasoning, The Natural Religion stresses our practical needs that are essential for our survival, but it also stresses that our mental experiences are our reality. Our emotions and feelings are who we are. Our material resources and our knowledge combine with our emotions and feelings to form our consciousness.

## 10.2: Development of our consciousness

It is difficult to identify one point as the start of our consciousness, it emerges gradually as our brain develops from when we are a foetus, on into the baby stage and childhood. It is made up of many complex parts and these parts develop at different times, influenced by our nutritional as well as stimulatory inputs. It is through our brain that our personality comes to the fore. Therefore, the mental stimulation we receive during our early development has a great effect on our feelings and personality later on in our development. It has been found that if as babies we have caring parents, who spend a significant amount of time playing with us and give us loving attention, then we grow up as more confident and reassured adults. A lack of love and affection as a baby and young child has been reported as causing an increased amount of anxiety, worry, and lack of confidence when we are teenagers and adults. The same trends have been seen in other animal species, such as monkeys and apes. These observations indicate that, as babies and children, we have important mental needs that influence how we form relationships with other people. Feeling loved and cared for, particularly when our brain is developing and growing, shapes our future consciousness.

All during our childhood and up to the stage that we are adults, our consciousness is forming. As we grow up, our capacity to understand the world around us also grows. Our emotional reactions, our feelings, and our control over them change. The relationships that children have with each other and with adults, also change as we grow up. For example, one area of development in our consciousness is that of our human sexuality. Our bodies develop and become capable of reproducing. This means we develop sexual feelings and drives that were not there before. Deep emotional feelings about sex become part of our consciousness as we grow towards adulthood. As adults, our consciousness embraces further feelings and emotions such as close personal relationships, and parental love and feelings of responsibility for children. With our changing role in life as we grow older, our consciousness also changes to suit that role. As teenagers and young adults we do not have the experience yet to put these changes of our consciousness in a wider context, or to fully assess their significance. As part of our preparation for adulthood, we can benefit from information that helps us to deal with this. Information and education about our mental needs and reactions help us to understand that our mental state and emotional requirements of our consciousness change as our life progresses.

The view included in The Natural Religion is that our consciousness develops gradually from the time we are born on into adulthood. It is not really possible to identify one point at which our consciousness starts. Our consciousness is not an indication of anything supernatural in our body, it is something that we feel in our brain, and that develops and changes from the point that we start life and continues to change all during our life.

## 10.3: **Our social emotional needs**

Being social animals, our brain has evolved feelings and emotions that suit and support our social activity and relationships. We love or hate, help or ignore, and decide whether to cooperate with or oppose people. Our consciousness has to strike the balance in our social relations with others – we are designed to deal with other people. Our human ecology is currently, and has always been, very much dependent on cooperation amongst people. Most of our achievements, such as increased food production and our technological upsurge, could not have been carried out without the coordinated efforts of many people. **We could not support over 6.6 billion people on this planet without our social cooperation and integration of the myriad of our specialized tasks. We need each other.**

Our brain and our consciousness have evolved to enable us to take part in social interaction. We are programmed to be in the company of other people. As mentioned before, one of the worst mental tortures to subject a person to is to stop them from having any contact with other people.

On the other hand, humans also compete with each other. Children in a family argue and fight over food, toys, and clothes. Parents compromise with each other about time spent on personal activities, as opposed to their family activities. Private businesses compete for customers, and people compete with each other for work. While ideally we should share all our resources, many resources are limited and therefore we have to compete for them. Yet, while we compete with people for certain resources, we also cooperate with these same people. It is our brain that takes all our, at times conflicting, circumstances into account and comes to a decision on how to act. We need social skills to reconcile our own interests with those of others, so we stay on good terms with the people we know. We need to be able to form friendships and social relationships so others will cooperate and share resources with us. Our complex social relationships mean that at the same time we cooperate and compete with, feel attracted towards and distance ourselves from, and depend on and help other people. Our brain needs to take all these various, and sometimes opposing, factors into account and decide on the best course of action. We are continuously conscious of our social position with respect to the people around us. **Our social lives therefore take a lot of mental and emotional effort, and occupies a lot of our consciousness.**

One of the most important social emotional needs that we have is to feel that we are liked. The fact that we feel that other people love and appreciate us is an important emotional need for us. This also has practical implications, because if we are liked that means we may be able to expect help and cooperation in one form or another. However, although all of us need help at some stage, just to have the feeling of being liked or loved is, for people the world over, an important emotional need in itself. Relationships and love are central to most people's lives. The company and the caring attention of a person has the power to satisfy our emotional needs, it makes us feel happy and contented. The full significance of the emotional needs that we receive from a relationship with a person is shown by the grief we feel when that person is no longer with us. A lack of friendship can have detrimental effects on our emotional health, as well as the health of the rest of our body. A lack of loving care when we are babies also influences our future personality. Our evolution as a social species has given us the emotional need for human company, friendship, and affection.

Besides love and affection, it is also important to most people that other people think well of us, respect us, and appreciate us. We like to feel accepted by others and seek their approval. We try to satisfy these emotional needs in a number of different ways. We try to gain recognition for our achievements from our family, friends, and acquaintances by achieving goals that we know will be appreciated by them. Many people also gather resources, such as land and money, in order to gain the respect of other people. Many choose a course in life that will give them a high social status, but not necessarily the highest monetary reward, that they potentially could achieve. The approval of others has a profound influence on the decisions we take in our lives. To be popular amongst one's peers is another very strong emotional need, for example, amongst teenagers. To gain this type of social approval is effectively an attempt to take as high a position in the social hierarchy as one can. Adults will also often attempt to ascend their social hierarchy by moving up in socioeconomic class. Seeking approval and social acceptance is part of our consciousness.

Forms of politeness and good manners are found in cultures and societies all over the world. These conventions can be very important in the social life of communities. Our emotional need to be treated politely and kindly is so important to us, that we generally recognize that to be polite means that we may not always be completely truthful. Yet we still prefer to have a slightly dishonest form of interacting with each other as long as it is respectful, kind, and sensitive to the other person's feelings and emotional needs. Politeness is so much part of our social life that to be impolite is generally disapproved of, and a transgressor may be subtly, or sometimes not so subtly, disciplined.

Our consciousness of how we are regarded by others is so much part of our everyday lives that we hardly realize how much it influences us. The way we speak, the way we act, our type of clothes, the way we wear our hair, our forms of politeness, are all designed to influence other people's opinions of us. Whole sections of our economy are concerned with fashions, which exist because we want to create a favourable impression with other people. Our emotional need for social approval and acceptance is very strong.

Our human social interaction includes competition, and this competition can at times take the form of violent conflict and war. When a society prepares for, or is at war, the information that it gives to both its civilian population as well as its armies is that the enemy are dangerous and offensive people. The enemy are shown in a light that makes them appear unworthy of our sympathies. Historically in human conflicts we try to dehumanize the enemy, so that the soldiers and the general public don't feel the same compassion for them as we would normally feel for other human beings. Political and military leaderships try to create an esprit de corps and feeling of togetherness within an army and a society, but they try to stop any of our social emotional needs from extending to the enemy – they try to create an 'us against them' feeling. This vilification and dehumanization of the enemy serves to counteract our normal social emotional inclinations to cooperate with others and form friendships. To be able to fight with each other we have to counteract our social feelings towards each other.

While we are emotionally designed to lead social lives and cooperate with others, we can also feel restricted by too much interaction with others. There is a limit to our need to interact with other people and if we experience too much social stimulation we can feel stressed, restricted, and confined by it. For instance, living in high population densities can limit one's choices, as we need strict regulation in order to cater for large numbers of people. We do cooperate with other people and we do need other people's company, but we also seek a certain amount of freedom. The popularity of 'getting away from it all' amongst people who can afford to do so, is an indication of our need for a certain amount of freedom. The fact that we can feel restricted and feel that we need more freedom means that our emotional needs, just like other aspects of our lives, operate between two extremes. We find that both having too little and having too much social contact is uncomfortable, so we seek a balance in between these two extremes that satisfies our social emotional needs.

Our emotions are stirred by someone who is in need of help – we feel we need to try to alleviate other people's suffering. To help others when it is likely that they will be in a position to return the favour in the future, makes practical sense for the person who first gives the help, but we also help people who are unlikely to be able help us in return. Many people who need help such as the sick, the poor, and the elderly may never be able to give help in return, yet most of our cultures and societies help them in one form or another. Parents have an inbuilt urge to care for their children and this is understandable for a number of reasons, one of which is that these children represent the genetic survival and continuation of the parents. However, people also care for others who are not related to them. In one survey carried out in Britain it was reported that as an activity, voluntary work was ranked only second to dancing for bringing joy. Many societies have organized health care systems that are free for those who need it and for which society at large pays, and many religions help people in need in various ways. As individuals we also help and care for people within and outside our families, for instance when they are sick. In agricultural society, particularly in time of surplus, people give food and other products to those who need them. **People living in affluent societies often give money to organizations who help people living in poorer regions. Helping people in need is part of our human ecology and satisfies part of our social emotional needs.**

As we have just seen in this section, and also included in The Natural Religion, is the viewpoint that not only is our social life central to our existence for practical reasons, but that we also need different types of contact with other people for our emotional needs. Our social emotional needs help to cement together the social framework of our lives. Our consciousness is very much occupied with striking balances in our social lives and assessing our position with respect to others.

Considering our emotional needs from a personal point of view, The Natural Religion draws attention to our feelings of compassion for others and our wish to help them. Our evolutionary history as social animals has instilled an instinctive reaction in our consciousness to help those in need. In other words, it feels good when we help others less fortunate than ourselves.

**One aim of this book is to help improve the quality of life of human beings. It therefore strongly advises all those who are in a position to help others to do so, as this is an immensely satisfying and emotionally rewarding experience. The Natural Religion, itself a product of the human consciousness, has these aims because of our need to help others.**

## 10.4: The Natural Religion – a product and a need of our consciousness

Our consciousness is who we are; it creates our world. It tells us we exist; it tells us that we were born and that we will die. Our consciousness, as far back as history records, has attempted to form a coherent view of our human existence. It has sought to explain our strong emotion for self-preservation even though we know we will die. It has sought to explain our basic drive to care for our own interests, while at the same time we also want to help other people. It has sought to explain our range of, at times seemingly contradictory, emotions that give us our experience of life. We as a species of animal have a large range of emotions and feelings that control our state of mind, and because we are self-aware we understand that we have these feelings. **Of all life on Earth we are, as far as we know, capable of the most objective understanding of ourselves; our consciousness needs to understand our consciousness.** In our attempts to achieve a coherent understanding of our own existence, we humans have formulated many possible explanations. People have formed religions to give coherence to sets of explanations about our existence. We also attempt to reconcile apparently opposing and confusing aspects of our lives to try to identify the best course of action for the future. Religions are one way by which our consciousness tries to satisfy its own needs. The need to know is a basic human emotional need and the subjects of our inbuilt curiosity very much includes ourselves. Therefore all religions, including The Natural Religion, are a product of the need of our consciousness to know and understand.

Philosophies also explore the extent of our knowledge. This differs from The Natural Religion in that it, like other religions, deals with our whole consciousness, including our feelings and emotions. **The philosophical approach focuses on knowledge information, and its meaning and relevance, in a more theoretical and academic way. However, The Natural Religion is a religion rather than a philosophy because its function is to help, inform, and support individual human beings with the practicalities of their everyday personal lives; it deals with real personal experience, it takes our whole consciousness into account.**

As we discussed in the beginning of this book, most current religions were formed in the past when much of the knowledge we now have, about ourselves and the world around us, was not yet available. So, a proportion of the ideas in traditional religions were based on guesses rather than on knowledge supported by evidence. The inclusion of mystical and supernatural ideas in traditional religions actually acknowledges the lack of information, as they are, by definition, beyond the natural world.

The Natural Religion suggests a new approach to religion and, as also already mentioned in the first chapters, has an advantage over older religions in benefiting (without having to resort to guesswork) from the huge upsurge in modern human knowledge. As pointed out in the Appendix and as with other religions, it is possible that the suggestions in this book may not all be correct.

But, by being based on up-to-date knowledge they aim to be more correct and create a better understanding of our human existence, and thereby satisfy people's need to know better, than the older religions. Each human being is different, and so our individual emotional needs also vary from person to person. Since religions address our emotional needs, it is a matter of personal preference which religion an individual person finds most satisfying. The Natural Religion recognizes different personal preferences, and advises that people inform themselves about the ideas of various religions and then choose the ideas they find most satisfying for their particular mental and emotional needs. Being based exclusively on knowledge, the approach in this book will not satisfy everybody. Many people may prefer mystical or supernatural interpretations of our existence, but our knowledge has now progressed to the stage that it can answer most questions about our existence, or indicate in which direction the answers lie. **This book, like other religions, is a product of our consciousness, created by it to satisfy its own need for understanding, this is the way our human brain works.** One aim of The Natural Religion is therefore that being based on up to date knowledge it can satisfy our human consciousness's need to know better than other more traditional religions.

A knowledge revolution has taken place in recent human history, and it is our combined human consciousness that has absorbed all of this knowledge. While our human consciousness has gained much new factual information and understanding in a relatively short period of time, our feelings and emotions have not undergone a similar revolution. Has our consciousness with its apparent current preoccupation with factual knowledge ignored its own emotional needs? Is the emotional side of our consciousness more controlled by our genes, and, therefore, not able to change as fast as our knowledge information?

Our consciousness is the way we experience our being, and both our emotions as well as our knowledge are part of that. So, now our feelings have to deal with a huge increase in knowledge and a deeper understanding of human life, but this can actually help our consciousness to reach a more coherent and reliable view of ourselves. More information about ourselves and the environment that created us, allows us to understand ourselves better. This knowledge information has been used in this book to construct The Natural Religion to help, amongst others things, to bring about a greater appreciation and more holistic recognition of the importance of emotions and feelings in our consciousness.

Religions help people to make decisions. As The Natural Religion is based on knowledge including human ecology, it takes into account that we have to strike balances between extremes in many aspects of our lives, and points out the, often unstable or undesirable, extremes of the options available to us. As we have seen in previous chapters, this book cannot determine the exact right balance for each person in each situation, but it can help to clarify the pros and cons of an issue, and each person then has to decide how much relevance these have in their own particular circumstances. The Natural Religion tries to inform about the advantages and disadvantages associated with each extreme, and how a course of action can reach a balance that is appropriate to a person's needs at any particular time. Part of this is that we stay sensitive to changes in our circumstances as time goes by, as we may need to shift our balances in response to changes to continue to have the outcome that we want.

**This new approach to religion tries to satisfy our consciousness's need to know, as well as help it to reach decisions about the best course of action for the future, and to assess them as time goes by. All of these functions are carried out by our consciousness, and it is The Natural Religion's role to help us do this.**

Our consciousness, as we know, also has social needs. By drawing attention to our shared human consciousness, and the extent to which we globally depend on each other, it is The Natural Religion's objective to help form relationships worldwide, and humanize our views of each other. This book uses our human knowledge to suggest how, and by what practical action, we can influence our futures for the benefit of ourselves and our species – this is effectively suggesting a set of ideals, or ethical and moral standards. Most societies have codes of ethics and moral rules, and these are generally based on its culture and religion. These ethical codes have developed over time in different parts of the world and while having many points in common, they also vary. Some cultures stress certain areas of human behaviour, while others deem other aspects to be more important. The underlying reasons for these differences may be due to circumstances at the times when these rules and codes were first laid down. The moral code that a particular person accepts and feels is the correct way to behave, is strongly influenced by the culture in which we grew up as a child. Moral codes and cultural values differ all over the world, so, up to the present, there is no global ethic or universal moral code that is the same for all humans. This does give rise to disagreements, mutual misunderstandings, distrust, and even conflict. **As part of creating a greater understanding and humanized view of each other worldwide, The Natural Religion seeks to contribute towards creating a global ethic or idealism that can be accepted by people from different cultures all over the world.**

Cultural and traditional religious values differ, so it is difficult to use these to form an ethic that is acceptable and applicable worldwide. On the other hand, like all the ideas incorporated in The Natural Religion, its ethical and moral values are based on our human knowledge. Knowledge information can be checked, and in whatever part of the world it is examined, it should lead to the same conclusions. This is one of the main criteria which ensures that our knowledge is reliable. **The conclusion reached in The Natural Religion is therefore that the greatest chance of achieving a global ethic is to base it on something that is the same all over the world. In other words, a global ethic based on our knowledge that is the same for, and can be checked by, all human beings.** By contributing towards creating a greater understanding of each other, the ideas in this book hope to strengthen our worldwide relationships, encourage peaceful means of resolving conflict, and also help towards the creation of a truly global ethic.

Therefore, by contributing towards achieving a global ethic, The Natural Religion hopes to help towards achieving the ideals of our global human consciousness.

Religions are a part of our human ecology, they are part of the way our consciousness works. This is shown by the fact that people all over the world have developed a succession and variety of religions. We search for a way to explain our knowledge and our emotional feelings and reactions to ourselves, so that it all makes sense and is satisfying to our consciousness. We know that our loved ones will die, and we know we ourselves will die, how do we feel about that? We need to care for our own interests, but we also need to care for others, what should we do? Is there a reason for our existence? Is there any sense to making plans for our future generations, since we ourselves are not going to be alive by then? There are many aspects of our existence as a species, that our consciousness attempts to reconcile and make sense of. Our consciousness has a need to try to take all aspects of our human life into account, and understand how all these factors combine together to result in the existence of an individual person, as well as our whole human species. As we have seen, it is a characteristic of our consciousness to attempt to arrive at an explanation that includes all factors and facets of human life. The set of ideas that forms the basis of religions are attempts at an all-inclusive explanation which reveals the reason for our existence and satisfies the need of people at large to understand.

The Natural Religion tries to show the function of apparently conflicting feelings and emotions. It attempts to reconcile our inbuilt emotional reactions to the changes in our human ecology due to the revolution in knowledge. As we know, The Natural Religion has an advantage over other religions, as it can consult a far greater body of knowledge than older more traditional religions had at their disposal when they were formed. Another difference between the time other religions were formed and the present day is, that our present vastly increased store of knowledge is also more easily available to a greater percentage of our population. Education has improved out of all recognition, particularly in richer regions of the world. It is particularly in the regions where people are better educated and can access knowledge more easily that more traditional religions, with beliefs held on faith, are being taken less seriously. Belief and faith tend to lose their effectiveness to satisfy the needs of our consciousness, if alternative explanations based on verifiable knowledge are easily available. This shows the importance of knowing and understanding to our consciousness. The existence of our older religions still demonstrate that all during our history, our human consciousness had the need to question and search for answers, even though we had much less knowledge then. Our human consciousness has always tried to satisfy its need for knowledge and it has always created religions, including The Natural Religion, as part of that effort. **The Natural Religion is in a better position to satisfy our need for our consciousness to know and understand, and by using this knowledge aims to achieve a good quality of life for all people worldwide.**

# Appendix

This Appendix discusses the two sets of terms: **knowledge information** and **genetic information**, and, **fact** and **faith**. While these words have generally accepted meanings, their exact definitions are of special significance in this book.

## Knowledge information and genetic information

The words 'knowledge' and 'information' are often used to express the same meaning. In the context of human ecology there is a difference between them because **we humans actually possess two types of information, namely knowledge information and genetic information.**

Our individually unique **genetic information** is contained in a chemical code within our genes. It is with us from the point, at the very beginning of our individual existence, when the genes of the sperm and egg combine until our death. Genetic information of a whole species evolves by means of natural selection and takes place over time during the course of many generations, but the genetic information of an individual living being is decided at the point of fertilization of the egg by the sperm, when a random combination of both parents' genes comes together. This means that human genetic information changes at random and only once per generation, but once an individual has it, it does not change during the normal course of a lifetime.

**Knowledge information** on the other hand has to be learned at various stages during the lifetime of each individual; we do not receive it when the egg from which we develop is fertilized. Unlike genetic information, we can continuously change and test and develop knowledge information during our entire lifetime. Therefore knowledge information is much more flexible and adaptable compared to genetic information.

One disadvantage that knowledge information has compared to genetic information is that it can be forgotten and lost by individuals as well as by whole societies. We don't know most of the knowledge that prehistoric humankind had because they did not write it down. Wars have ended whole cultures along with much of their knowledge. Entire libraries of books have been destroyed, for example the library in Alexandria, which was then the biggest library in the world, was burned by Archbishop Theophilus in the late 4th century. Our powerful position as human beings on this Earth is mostly due to our capability of having knowledge; but if we were to lose much of it, our lives would become like the lives of our prehistoric ancestors. Our genetic information, on the other hand, will always be part of us as long as we humans exist.

### What is knowledge made of?

While we all know what it is like to be conscious, to think, and to have memories, it is difficult to imagine what thoughts are actually made of. When our brain was not as well understood as it is now, our consciousness and self-awareness was used by various religions as an indication of something supernatural within us. This is understandable given that we used not to have any knowledge of the structure and functioning of our brain. Our consciousness is the way we experience reality, it is our personality and the essence of ourselves, we don't experience our consciousness as something material like we do our hands and feet and other parts of our body. It feels immaterial to us, weightless and without substance, an ethereal entity. So, it is not surprising that our thoughts and consciousness were interpreted as our ethereal spirit, our supernatural part. However, we now know the structure of brain tissue, how it connects with all our senses, and how it controls our muscles. We know why our brain feels non-physical, and we can now account for our thoughts, consciousness, and self-awareness without any supernatural explanations.

There are two basic aspects to our human brain that enable it to perform its amazing function. The first is, that it is incredibly complex on a microscopic level. And the second is, the combination of chemical and electrical activity in the brain.

The complexity of the human brain is truly awesome. It is one of the most complex organs in the animal kingdom and consists of up to 100 billion brain cells, each of which has between 10,000 and 100,000 contact points with other brain cells. To help us to visualize such huge numbers: there are about as many nerve cells in the human brain as there are trees in the Amazonian rainforest. Even more amazing is, that the number of contacts between all the cells in our brain is similar to all the leaves on all the trees in the Amazonian rainforest. All of these 100 billion cells – with all their tens of thousands of points of contact each – are contained within one human brain. When we realize the sheer magnitude of the number of cells, multiplied by the contact points between them, then we begin to understand the level of thought, feeling and emotion of which the human brain is capable.

Knowledge information itself is held by our 100 billion brain cells in communication with each other via their tens of thousands of contact points. They do this by a combination of chemical reactions and tiny electrical impulses. This chemical and electrical activity is influenced by the information that the nerve cells from our senses send to our brain. The electrical and chemical activity also influences how our brain cells and their contact points grow. So our knowledge exists as chemical and electrical activity which is effectively a type of code, as well as in the connections that our brain cells make with each other. This electrochemical activity together with our brain cell structure is our perception of our life and the world we live in, this is our experience and learning, our intuition and understanding, our feelings and emotions, our thoughts and awareness of reality around us. This is the way we store our knowledge information, this is our consciousness.

Our brain allows us to experience and influence happenings ranging from lifting a cup to our lips so we can drink from it without spilling it all over ourselves, to having relationships with other people, enjoying art, landing people on the moon, and exploring outer space.

Our human brain is as far as we know the most developed brain which exists and is amazing both in terms of its structure and its capabilities. It is the reason why we humans have gained so much more control over the world around us than other animal species.

### Storing and retrieving knowledge information

A very important aspect of knowledge information is that, besides using our brain, we have found other ways to store and retrieve it. One way in which we can do this is by using **writing**. Once we can read the writing, we can retrieve the knowledge information that the writing represents and take it back into our brains again. Knowledge information can be given from one person to another without the person imparting it ever meeting the person receiving it. In fact, by writing, a person can still communicate knowledge many years, indeed centuries, after they have died. Knowledge information can also of course be stored and transferred by means of other techniques. Bunches of ropes knotted in a certain way, notches cut into the edges of stones, and symbols carved into stone, wood and metal have all been used to store knowledge and communicate this from earliest times.

Most recently we have developed methods of storage and retrieval of knowledge far beyond these humble beginnings. With electronic storage and retrieval of knowledge information we can handle huge amounts of data using computers, and this has revolutionized the availability of knowledge information to humankind. The Internet represents the biggest library in human history. And it can be accessed by any computer from any location around the world that has a connection to the Internet. The ease of access of such a gigantic amount of information is unprecedented and could have a great influence on our global human ecology.

### Knowledge information needs to be learned

Knowledge is not genetically inherited, so we have to learn it. To facilitate this necessary process, various systems of **learning and education** have been developed. Education can occur formally in schools and/or informally through learning from others in everyday life, and is essential for every person. Societies and cultures would disintegrate without it, our current human ecology is shaped by knowledge, all of which has to be learned.

As mentioned before, a weakness of knowledge information is that it can be lost. To keep it, it needs to be learned again and again by each succeeding generation. If we lost much of our agricultural, health care, and construction knowledge we would lose the capacity to feed ourselves, keep ourselves healthy, and provide shelter for ourselves.

## Fact and faith

This book offers a new approach to religion based on **facts** rather than on **faith**. So we will now take a closer look at precisely what facts and faith are, and how they differ.

### Facts

Facts and faith are fundamentally different. Facts are gathered and built up by checking and testing theories and information. They need to be able to satisfy criteria, such as whether they can stand up to examination and questioning and/or reliably predict future events. We need to predict future events on a daily basis ranging from crossing the street without getting run over and drinking from a cup without spilling the content, to timing the planting of crops so we get a good harvest and designing buildings that don't fall down. Facts allow us to reliably achieve all these goals.

However, as reliable as most of our facts have proven to be, some of them, thought to be correct at one time, were subsequently found to be incorrect. So, it is likely that not all our present-day facts are correct either, and therefore might need to be adapted or changed. By questioning and checking our facts, we constantly verify our knowledge information and test its dependability, and also continuously accumulate new facts. While as a species we currently have more reliable information and facts at our disposal than we ever had before, the process of accumulating and refining facts is not finished. We don't know if it will ever be finished. Some of the ideas presented in this or any book that is based on a developing and ever-changing body of knowledge or facts, may have to be adapted, altered, corrected, or further developed in the future. However, the facts on which this book is based did come about by means of a rigorous process of questioning and checking, therefore they do have a high degree of reliability and, of course, continue to be open to verification and checking.

Another aspect of factual information is that a fact is only reliable if it is the same wherever in the world it is checked. Facts are not dependent on culture, society, tradition, or fashion. For instance, the language that scientists use varies depending on where in the world they are, but the facts and findings that they are talking about are the same. Where you are born does not influence facts.

### Faith

In contrast, faith is formed and maintained in a very different manner to facts. Faith means that certain ideas are accepted without question. Faith in super-natural beliefs such as the existence of gods, the wishes or demands of gods, the afterlife and other spiritual concepts cannot be checked. Being super-natural – or beyond nature – they, by definition, cannot be physically tested or verified. Since it is not possible to check supernatural beliefs, they have to be accepted by believers on faith. Many millions of people accept ideas on faith. Most religions exhort believers to accept a whole body of religious belief on faith and advocate that they should regard it as the truth. However being based on faith this cannot be proven; but for the same reason that supernatural beliefs cannot be proven, they cannot be disproven either.

Which faith a person believes in is clearly very much influenced by where in the world the believer was born and of which culture they are part. A person born in Indonesia is 95% likely to be a Muslim, while most people born in Ireland are Christians. As mentioned before, faiths differ all over the world and because they contradict each other they cannot all be correct, but it is not possible to check which one is right because faiths deal with supernatural ideas.

A general comparison of fact and faith is as follows. Fact is based exclusively on nature and the world around us, while faith usually involves beliefs in supernatural concepts. In terms of reliability, facts are based on much more stringent and robust criteria than faith, facts can and should be tested and rechecked and verified, faith cannot be tested. Reliable facts are the same all over the world, while the world's faiths differ and contradict each other. It is therefore irresponsible to base important decisions, conclusions and suggestions about what people should do on faith, particularly when facts are available. This is the reason why the advances in religion suggested in this book are based on facts and not on faith. A fact religion is not only more reliable and responsible than faith religions, but is also applicable worldwide. This religion based on fact, The Natural Religion, has a greater chance of achieving a truly global ethic and moral ideals than faith religions with their varying and contradicting beliefs.

# Bibliographic Sources

## Sources include:

Aitchison, J., 1996. *The Seeds of Speech – Language Origin and Evolution.* Cambridge: Cambridge University Press.

Alcock, J., 1993. *Animal Behavior – An Evolutionary Approach.* (5th edition) Sunderland, Massachussetts: Sinauer Associates.

Allaby, M., 1989. *Guide to Gaia.* London: MacDonald Optima.

Anonymous, Online: *www.cia.gov/library/publications/the-world-factbook/print/xx.html.*

Anonymous, Online: *www.census.gov/main/www/popclock.html.*

Appleyard, B., 1992. *Understanding the Present.* London: Picador .

Appleyard, B., 1996. Origin of Species. *The Sunday Times,* 12 May 1996, 7.1–7.2.

Ardrey, R., 1977. *The Hunting Hypothesis.* Fontana/Collins.

Arnett, B., 2000. The Sun. Online: *www.seds.org/nineplanets/nineplanets/ sol.html: SEDS, Students for the Exploration and Development of Space.*

Attenborough, D., 1981. *Life on Earth.* Glasgow: Fontana/Collins.

Attenborough, D., 1991. *Overleven in de Natuur* (transl.: Kaat, J.). Baarn: Bosch & Keuning.

BPC Publishing, 1975. *Encyclopedia of World Religions.* London: Octopus Books.

Baigent, M., Leigh, R. & Lincoln, H., 1983. *The Holy Blood and The Holy Grail.* London: Corgi Books, Transworld Publishers.

Baker, R. R., 1996. *Sperm Wars – Infidelity, Sexual Conflict and Other Bedroom Battles.* London: Fourth Estate.

Baker, R. R. & Oram, E., 1999. *Baby Wars – The Dynamics of Family Conflict.* Hopewell, New Jersey: The Ecco Press.

Baker, R. R., 1999. *Sex in the Future – Ancient Urges Meet Future Technology.* London: Macmillan.

Beaver, R. P., Bergman, J., Langley, M.S., Metz, W., Romarheim, A., Walls, A., Withycombe, R. & Wootton, R.W.T. (Ed.), 1988. *The World's Religions.*Oxford: Lion Publishing.

Bodmer, W. & McKie, R., 1994. *The Book Of Man – The Quest to Discover Our Genetic Heritage.* London: Abacus.

Bowker, J. (Ed.), 1997. *The Oxford Dictionary of World Religions.* Oxford: Oxford University Press.

Boyden, S., Millar, S., Newcombe, K. & O'Neill, B., 1981. *The Ecology of a City and its People – The Case of Hong Kong.* Canberra: Australian National University Press.

Boyden, S., Dovers, S. & Shirlow, M., 1990. *Our Biosphere Under Threat – Ecological Realities and Australia's Opportunities.* Melbourne: Oxford University Press Australia.

Bronowski, J., 1973. *The Ascent of Man.* London: Futura.

Brooke, J. H., 1991. *Science and Religion – Some Historical Perspectives.* Cambridge: Cambridge University Press.

Brown, L. R., Flavin, C. & Starke, L. (Ed.), 1996. *State of the World – A Worldwatch Institute Report on Progress Toward a Sustainable Society.* New York: W.W. Norton.

Bryson, B., 2004. *A Short History of Nearly Everything.* London: Black Swan.

Budiansky, S., 1995. *Nature's Keepers – The New Science of Nature Management.* London: Weidenfeld & Nicholson.

Burley, J. (Ed.), 1999. *The Genetic Revolution and Human Rights – The Oxford Amnesty Lectures 1998.* Oxford: Oxford University Press.

Byrne, R., 1995. *The Thinking Ape – Evolutionary Origins of Intelligence.* Oxford: Oxford University Press.

Cairns-Smith, A. G., 1990. *Seven Clues to the Origin of Life.* Cambridge: Cambridge University Press.

Campbell, B., 1983. *Human Ecology – The Story of our Place in Nature from Prehistory to the Present.* London: Heinemann Educational.

Carson, R. L., 1963. *The Sea Around Us.* New York: Signet.

Carson, R. L., 1965. *Silent Spring.* Harmondsworth: Penguin Books.

Carter, R., 2000. *Mapping the Mind.* London: Phoenix.

Cavalli-Sforza, L. L. & Cavalli-Sforza, F., 1995. *The Great Human Diaspora – The History of Diversity and Evolution* (transl.: Thorn, S.). Reading, Massachusetts: Helix Books, Addison-Wesley.

Chandler, D., 2000. *Bosnia – Faking Democracy After Dayton.* (2nd edn London: Pluto Press.

Chardin, P. T. de, 1959. *The Phenomenon of Man* (transl.: Wall, B.). London: Collins.

Chardin, P. T. de, 1964. *The Future of Man* (transl.: Denny, N.). New York: Harper Torchbooks, Harper & Row.

Chardin, P. T. de, 1966. *Man's Place in Nature – The Human Zoological Group* (transl.: Hague, R.). New York: Harper & Row.

Chopra, D., 2000. *How To Know God – The Soul's Journey Into The Mystery Of Mysteries.* London: Rider Books, Random House.

Claessen, H. J. M., 1974. *Politieke Antropologie – Een Terreinverkenning.* Assen: Van Gorcum.

Clark, G., 1994. *Space, Time and Man – A Prehistorian's View.* Cambridge: Cambridge University Press.

Cohen, J. E., 1995. *How Many People Can the Earth Support?* New York: W.W. Norton.

Coleman, S. & Watson, H., 1992. *An Introduction to Anthropology.* London: Tiger Books.

Colinvaux, P., 1980. *Why Big Fierce Animal are Rare.* London: Penguin Books.

Colinvaux, P., 1986. *Ecology.* New York: John Wiley & Sons.

Conn, E. E. & Stumpf, P. K., 1972. *Outlines of Biochemistry*. (3rd edition) New York: John Wiley & Sons.

Connolly, B., 1997. *Traditional Fishery Knowledge and Practice for Sustainable Marine Resources Management in Northwestern Europe – A Comparative Study of Fisheries in Ireland and The Netherlands*. Research Report, Online: http://homepage.eircom.net/~eufisheries/

Connolly, B., 1998. Human ecology: one coherent unit, not a multi- or inter-disciplinary amalgam. *Journal of Human Ecology, 9*(4), 297–310.

Connolly, B., 2001. Recommendations of appropriate systems of sea tenure for future fisheries management. *Traditional Marine Resource Management and Knowledge Information Bulletin, 13*, 24–27.

Connolly, P., Gonzalo, F., Forsyth, J., Fowler, P. & Twyford, P. (Ed.), 2002. *Rigged Rules and Standards – Trade, Globalisation, and the Fight against Poverty*. Oxfam International.

Cooper, D. E. & Palmer, J. A. (Ed.), 1998. *Spirit of the Environment – Religion, Value and Environmental Concern*. London: Routledge.

Crick, F., 1995. *The Astonishing Hypothesis*. London: Simon & Schuster.

Daly, M. & Wilson, M., 1983. *Sex, Evolution and Behavior*. (2nd edition) Belmont, California: Wadworth Publishing.

Darling, D., 1995. *After Life – In Search of Cosmic Consciousness*. London: Fourth Estate.

Darwin, C., 1936. *The Origin of Species by Means of Natural Selection, or the Preservation of Favoured Races in the Struggle for Life*, and, *The Descent of Man and Selection in Relation to Sex*. New York: The Modern Library (Based on 1st edition, 1859, and 6th edition, 1872).

Darwin, C., 1979. *The Expression of Emotions in Man and Animals*. London: Julian Friedmann(Based on 1872 edition).

Davis, D. E., 1986. Regulation of Human Populations in Northern France and Adjacent Lands in the Middle Ages. *Human Ecology, 14* (2), 245–267.

Dawkins, R., 1976. *The Selfish Gene*. St. Albans: Granada Publishing. (Paladin Books).

Dawkins, R., 1982. *The Extended Phenotype*. Oxford: Oxford University Press.

Dawkins, R., 1986. *The Blind Watchmaker*. London: Penguin Books.

Dawkins, R., 1995. *River Out of Eden*. London: Phoenix.

Dawkins, R., 2006. *The God Delusion*. London: Bantam Press.

Diamond, J., 1991. *The Rise and Fall of the Third Chimpanzee*. London: Random House.

Diamond, J., 1998. *Guns, Germs and Steel – A Short History of Everybody for the Last 13,000 Years*. London: Random House.

Diesendorf, M. & Hamilton, C. (Ed.), 1997. *Human Ecology, Human Economy – Ideas for an Ecologically Sustainable Future*. St. Leonards, Australia: Allen & Unwin.

Donaldson, M., 1992. *Human Minds – An Exploration*. London: Penguin Books.

Douglas-Hamilton, I. & Douglas-Hamilton, O., 1975. *Among the Elephants*. London: Collins and Harvell.

Editorial Nature Biotechnology, 1999. Much more than elegant. *Nature Biotechnology, 17* (January), 1.

Eibl-Eibesfeldt, I., 1975. *Ethology – The Biology of Behavior.* (2nd edition)New York: Holt, Rinehart and Winston.

Eibl-Eibesfeldt, I., Kortlandt, A., Eisentraut, M., Krieg, H., Freye, H-A., Mohr, H.C.E., Grzimek, B., Piechocki, R., Hediger, H., Rahm, U., Heinemann, D., Slijper, E.J., Hemmer, H. & Thenius, E. (Ed.), 1975. *Grzimek's Animal Life Encyclopedia – Mammals II* (English edition). New York: Van Nostrand Reinhold Company.

Fiedler, W., Gewalt, W., Grzimek, B., Heinemann, D., Herter, K. & Thenius, E. (Ed.), 1972. *Grzimek's Animal Life Encyclopedia – Mammals I* (English edition). New York: Van Nostrand Reinhold.

Fossey, D., 1983. *Gorillas in the Mist.* Boston: Houghton Mifflin..

Franke, T., 1996. Dood gaan we allemaal. *Intermediaire*, 8 November 1996, 41–43.

Frazer, J. G., 1922. *The Golden Bough – A Study in Magic and Religion.* London: Papermac (1990 edition).

Galdikas, B. M. F., 1995. *Reflections of Eden – My Life with the Orangutans of Borneo.* London: Indigo.

Giddens, A., 1993. *Sociology.* Oxford: Polity (2nd edition).

Gjertsen, D., 1989. *Science and Philosophy.* London: Pelican Books.

Goodall, J., 1971. *In the Shadow of Man.* London: Phoenix (Revised edition 1996).

Goodall, J., 1991. *Oog in Oog met Chimpanzees – 30 Jaar in het Oerwoud van Gombe* (transl.: Jongh, T. de). Amsterdam: Uitgeverij Veen.

Goodall, J., 1994. Postscript – Conservation and the Future of Chimpanzees and Bonobos Research in Africa. In: Wrangham, R. W. (Ed.), *Chimpanzee Cultures*, (397–404). Cambridge, Massachusetts: Harvard University Press.

Gould, S. J., 1980. *The Panda's Thumb – More Reflections in Natural History.* London: Penguin Books.

Gould, S. J., 1981. *The Mismeasure of Man.* London: Penguin Books.

Graven, J., 1967. *Non-Human Thought – The Mysteries of the Animal Psyche* (transl.: Salemson, H. J.). New York: Stein and Day.

Greene, M. E. & Rao, V., 1995. The Marriage Squeeze and the Rise in Informal Marriage in Brazil. *Social Biology*, *42* (1–2), 65–82.

Greenfield, S., 1998. *The Human Brain – A Guided Tour.* London: Phoenix.

Gregory, D., Martin R. & Smith, G. (Ed.), 1994. *Human Geography – Society, Space and Social Science.* London: Macmillan Press.

Gribbin, M. & Gribbin, J., 1995. *Being Human – Putting People in an Evolutionary Perspective.* London: Phoenix.

Gribbin, J. 1995. Basic Instincts. *The Sunday Times*, 7 May 1995, 7.5.

Grzimek, H. C. B., Illies, J. & Klausewitz, W. (Ed.), 1976. *Grzimek's Encyclopedia of Ecology* (English edition). New York: Van Nostrand Reinhold.

Harding, S. (Ed.), 1993. *The "Racial" Economy of Science – Towards a Democratic Future.* Bloomington and Indianapolis: Indiana University Press.

Harris, M., 1995. *Culture, People, Nature – An Introduction to General Anthropology.* New York: HarperCollins College Publishers (6th edition).

Harrison, S., 1992. Ritual as intellectual property. *MAN, The Journal of the Royal Anthropological Institute*, *27* (2), 229–244.

Hawking, S. W., 1988. *A Brief History of Time – From the Big Bang to Black Holes.* London: Bantam Press.

Hayden, T., 1996. *The Lost Gospel of the Earth*. Dublin: Wolfhound Press.

Heberer, G. & Wendt, H. (Ed.), 1976. *Grzimek's Encyclopedia of Evolution* (English edition). New York: Van Nostrand Reinhold.

Hintum, M. van, 1996. Veel beroepsgebonden geslachtsverschillen berusten louter op traditie – Hersenonderzoeker Swaab relativeert stereotype taakverdeling tussen mannen en vrouwen. *Mare, Leids Universitair Weekblad*, 7 May 1995, 6.

Hölldobler, B. & Wilson, E. O., 1994. *Journey to the Ants – A Story of Scientific Exploration*. Cambridge, Massachussetts: The Belknap Press of Harvard University Press.

Hughes, J. D. & Thirgood, J.V., 1982. Deforestation in Ancient Greece & Rome – a cause of collapse. *The Ecologist, 12*(5), 196.

Hulspas, M., 1996. Een Roomse verleiding. *Intermediair*, 8 November 1996, 9.

Human Genome Project, 2000. Human Genome Project Information. Online: *www.ornl.gov/hgmis/*: *U.S. Department of Health*.

Immelman, K. (Ed.), 1977. *Grzimek's Encyclopedia of Ethology* (English edition). New York: Van Nostrand Reinhold.

Jacobs, M. & Stern, B. J., 1955. *General Anthropology*. New York: Barnes & Noble.

Jamieson, N. L. & Lovelace, G. W., 1985. Cultural Values and Human Ecology – Some Initial Considerations. In: Hutterer, K. L., Rambo, A. T. & Lovelace, G.W. (Ed.), *Cultural Values and Human Ecology in Southeast Asia*, (27–54). Michigan: The University of Michigan.

Janssen, J., 1996. Deze aap denkt als een mens. *Weekeinde*, 1 June 1996, 8–9.

Johanson, D. C. & Maitland, E. A., 1990. *Lucy – The Beginnings of Humankind*. London: Penguin Books.

Johnson, R.D., 2003. Homosexuality: Nature or Nurture. Online: *http://allpsych. com/journal/homosexuality.html*.

Karpov, S. P., 1993. The grain trade in the southern Black Sea – the thirteenth to the fifteenth century. *Mediterranean Historical Review, 8* (1), 55–73.

Kassam, A., 1996. The Booram Oromo Gadamojji Ceremony – Held at Sololo (Kenya) June-July 1995. *Culture & History, 1* (3), 14–39.

Krebs, J. R. & Davies, N.B., 1993. *An Introduction to Behavioural Ecology*. Oxford: Blackwell Science (3rd edition).

Krebs, C. J., 1994. *Ecology – The Experimental Analysis of Distribution and Abundance*. New York: HarperCollins College Publishers (4th edition).

Leakey, R. E. & Lewin, R., 1993. *Origins Reconsidered – In Search Of What Makes Us Human*. London: Abacus.

Leakey, R. E. & Slikkerveer, L. J., 1993. *Man-Ape Ape-Man – The Quest for Human's Place in Nature and Dubois' "Missing Link"*. Leiden: The Netherlands Foundation for Kenya Wildlife Service.

Leakey, R. E., 1994. *The Origin of Humankind*. London: Phoenix (Science Masters).

Leakey, R. E., 1996. *Human Origins*. London: Phoenix (Abridged edition).

Legesse, A., 1987. Oromo Democracy. Conference on: Oromo Revolution. Washington D.C.

Lewin, R., 1993. *Complexity – Life on the Edge of Chaos*. London: Phoenix (1995 impression).

Lewis, H. S., 1994. Aspects of Oromo political culture. *The Journal of Oromo Studies, 1*(1–2), 53–58.

Lewis, H. S., 1995. Democracy and Oromo Political Culture. *Life & Peace Review*, 4 , 26–29.

Lewontin, R. C., 1993. *The Doctrine of DNA – Biology as Ideology*. London: Penguin Books.

Lorenz, K., 1957. *Ik Sprak met Viervoeters, Vogels en Vissen* (transl.: Warren, H.). Amsterdam: Uitgeverij Ploegsma.

Lovelock, J., 1995. *Gaia – A New Look at Life on Earth*. Oxford: Oxford University Press (Reissued 1995).

Malthus, T. R., 1798 + 1830. *An Essay on the Principle of Population*, and, *A Summary View of the Principle of Population*. London: Penguin Books (1970 edition).

Manning, A. & Stamp Dawkins, M., 1992. *An Introduction to Animal Behaviour*. Cambridge: Cambridge University Press (4th edition).

Marsden, P., 2001. It is one of man's oldest battles: the hunter-gatherer vs. the farmer. *The Sunday Times Supplement*, 4 March 2001, 40–41.

Masson, J. & McCarthy, S., 1994. *When Elephants Weep – The Emotional Lives of Animals*. London: Vintage (Vintage paperback 1996).

Maynard Smith, J., 1988. *Did Darwin Get It Right – Essays on Games, Sex and Evolution*. London: Penguin Books.

Maynard Smith, J., 1993. *The Theory of Evolution*. Cambridge: Cambridge University Press.

McHenry, H. M., 1994. Behavioral ecological implications of early hominid body size. *Journal of Human Evolution*, 27 , 77–87.

McLaren, D., Bullock, S. & Nusrat, Y., 1998. *Tomorrow's World – Britain's Share in a Sustainable Future*. London: Earthscan Publications Limited.

McNeill, W. H., 1991. *The Rise of the West – A History of the Human Community, with retrospective essay*. Chicago: University of Chicago Press.

Milton, R., 1992. *The Facts of Life – Shattering the Myths of Darwinism*. London: Corgi (Reprint 1994).

Moore Lappé, F. & Shurman, R., 1995. The Population Debate. In: Kirkby, J., O'Keefe, P. & Timberlake, L. (Ed.), *The Earthscan Reader in Sustainable Development*, (104–109). London: Earthscan.

Morgan, E., 1995. *The Descent of the Child – Human Evolution from a New Perspective*. Oxford: Oxford University Press.

Morris, D., 1969. *The Human Zoo*. London: Jonathan Cape.

Murdoch, G. P., 1981. *Atlas of World Cultures*. Pittsburgh, PA.: University of Pittsburgh Press.

Myers, N. (Ed.), 1985. *The Gaia Atlas of Planet Management – For Today's Caretakers of Tomorrow's World*. London: Pan Books.

Myers, N. & Kent, J. (Ed.), 2005. *The New Gaia Atlas of Planet Management*. London: Gaia Books.

Nas, P. J. M., 1990. *De Stad in de Derde Wereld – Een Inleiding tot de Urbane Antropologie en Sociologie*. Muiderberg: Dick Coutinho.

Nas, P. J. M., 1993. Introduction (of Urban Symbolism). In: Nas, P. J. M. (Ed.), *Urban Symbolism*, (1–12). Leiden: E.J. Brill.

Nas, P. J. M., 1993. *Jakarta, City full of Symbols – An Essay in Symbolic Ecology*. In: Nas, P. J. M. (Ed.), *Urban Symbolism*, (13–37). Leiden: E.J. Brill.

O'Donohue, J., 1997. *Anam Cara – Spiritual Wisdom from the Celtic World.* London: Bantam Press.

Oliver, R., 1993. *The African Experience.* London: Pimlico.

Oliver, J. S., Sikes, N.E. & Stewart, K.M., 1994. Introduction to "early hominid behavioural ecology": new looks at old questions. *Journal of Human Evolution, 27,* 1–5.

Orlove, B. S., 1980. Ecological anthropology. *Annual Review of Anthropology, 9 ,* 235–273.

Pelto, P. J., 1965. *The Study of Anthropology.* Columbus, Ohio: Charles E. Merrill Books.

Pianka, E. R., 1966. Latitudinal gradients in species diversity – a review of concepts. *The American Naturalist, 100* (910), 33–46.

Póirtéir, C. (Ed.), 1995. *The Great Irish Famine.* Dublin: RTÉ/Mercier Press.

Pracht, M., 1996. *Geology of Dingle Bay.* Dublin: The Geological Survey of Ireland.

Quinn, H. & McDunn, R. A., 2000. Theory – Quarks. Online: *www2.slac.stanford. edu/vvc/theory/quarks.html: Stanford Linear Accelerator Center.*

Reader, J., 1990. *Man on Earth – A Celebration of Mankind.* New York: Harper & Row, Publishers.

Reeves, H., 1986. *De Giftige Steek van de Kennis – Over Zin en Ontwikkeling van het Heelal* (transl.: Rutten-Vonk, R.). Amsterdam: van Gennep.

Renerew, C., 1992. Archaeology, genetics and linguistic diversity. *MAN, The Journal of the Royal Anthropological Institute, 27* (3), 445–478.

Rensch, B., 1972. *Homo Sapiens – From Man to Demigod* (transl.: Sym, C.A.M.). London: Methuen.

Resnick, M., 1994. *Turtles, Termites and Traffic Jams – Explorations in Massively Parallel Microworlds.* Cambridge, Massachussetts: MIT Press.

Romer, A. S., 1954. *Man and the Vertebrates: 1.* Middlesex: Penguin Books.

Romer, A. S., 1954. *Man and the Vertebrates: 2.* Middlesex: Penguin Books.

Rowley, J. & Holmberg, J., 1995. Stabilizing population: the biggest challenge. In: Kirkby, J., O'Keefe, P. & Timberlake, L. (Ed.), *The Earthscan Reader in Sustainable Development,* (115–122). London: Earthscan .

Russell, B., 1952. *The Impact of Science on Society.* London: Unwin Hyman.

Russell, B., 1961. *History of Western Philosophy – And its Connection with Political and Social Circumstances from the Earliest Times to the Present Day.* London: Routledge.

Shaw, B. D., 1992. Explaining incest: brother-sister marriage in Graeco-Roman Egypt. *MAN, The Journal of the Royal Anthropological Institute, 27* (2), 267–299.

Shiva, M., 1995. The politics of population policies. In: Kirkby, J., O'Keefe, P. & Timberlake, L. (Ed.), *The Earthscan Reader in Sustainable Development,* (110–114). London: Earthscan.

Skinner, Q., 1985. *The Return of Grand Theory in the Human Sciences.* Cambridge: Cambridge University Press (Canto edition 1994).

Sponsel, L. E., 1985. Ecology, anthropology, and values in Amazonia. In: Hutterer, K. L., Rambo, A. T. & Lovelace, G. (Ed.), *Cultural Values and Human Ecology in Southeast Asia,* (77–122). Michigan: The University of Michigan.

Starke, L. (Ed.), 2006. *State of the World 2006 – A Worldwatch Institute Report on Progress Toward a Sustainable Society.* New York: W.W. Norton.

Stokes, B. J., 1967. *Organic Chemistry*. London: Edward Arnold (2nd edition).

Susman, R. L. (Ed.), 1984. *The Pygmy Chimpanzee – Evolutionary Biology and Behavior*. New York: Plenum Press.

Ta'a, T., 1996. Traditional and modern cooperatives among the Oromo. In: Baxter, P. T. W., Hultin, J. & Triulzi, A. (Ed.), *Being and Becoming Oromo – Historical and Anthropological Enquiries*, (202–209). Uppsala: Nordiska Afrikainstitutet.

Tambiah, S. J., 1990. *Magic, Science, Religion, and the Scope of Rationality*. Cambridge: Cambridge University Press.

Thio, A., 1989. *Sociology – An Introduction*. New York: Harper & Row, Publishers. (2nd edition).

Thomas, H., 1989. *An Unfinished History of the World*. London: Pan Books.

Thompson, R. F. (Ed.), 1972. *Physiological Psychology*. San Francisco: W.H. Freeman .

Vidal, G., 2000. *The Golden Age – A Novel*. London: Little, Brown.

Villee, C. A. & Detier, V. G., 1971. *Biological Principles and Processes*. Philadelphia: W.B. Saunders.

Vineberg, E. O., 1997. Pan paniscus/Bonobo News. *Pan paniscus/ Bonobo News*, 4(1).

Viney, M., 1995. Second-guessing the mind of God. *The Irish Times. Weekend*, 10 June 1995, 1.

Voeten, T., 1996. *Tunnelmensen*. Amsterdam: Atlas.

Waal, F. B. M. de, 1996. *Good Natured – The Origins of Right and Wrong in Humans and Other Animals*. Cambridge, Massachusetts: Harvard University Press.

Waal, F. B. M. de & Lanting, F., 1997. *Bonobo – The Forgotten Ape*. Berkeley: University of California Press.

Waal, F. B. M. de, 1998. *Chimpanzee Politics – Power and Sex amongst Apes*. Baltimore: The Johns Hopkins University Press.

Walker, T. & Aumiller, L., 1993. *River of Bears*. Shrewsbury: Swan Hill Press.

Watson, L., 1973. *Supernature – The Natural History of the Supernatural*. London: Hodder and Stoughton.

Watson, L., 1995. *Dark Nature – A Natural History of Evil*. London: Sceptre Paperback, Hodder and Stoughton.

Wills, C., 1995. *The Runaway Brain – The Evolution of Human Uniqueness*. London: Flamingo.

Wilson, E. O., 1975. *Sociobiology – The Abridged Edition*. Cambridge, Massachusetts: The Belknap Press of Harvard University Press.

Wilson, E. O., 1994. *The Diversity of Life*. London: Penguin Books.

Wilson, E. O., 1995. *On Human Nature*. London: Penguin Books.

Wilson, E. O., 1998. *Consilience – The Unity of Knowledge*. London: Abacus, Little, Brown.

Woodham-Smith, C., 1962. *The Great Hunger – The Horrific Story of the Irish Famine*. London: New EnglishLibrary.

Wright, R., 1993. Science, God and Man. *Time*, 4th January 1993, 44–48.

Ziman, J., 1976. *The Force of Knowledge – The Scientific Dimension of Society*. Cambridge: Cambridge University Press.

# Index

Printed in the United Kingdom
by Lightning Source UK Ltd.
135831UK00002B/106-183/P